Universitext

Giuseppe Da Prato

An Introduction to Infinite-Dimensional Analysis

 Springer

Giuseppe Da Prato
Scuola Normale Superiore
Piazza dei Cavalieri 7
56100 Pisa, Italy
e-mail: daprato@sns.it

Mathematics Subject Classification (2000): 37L55, 60H10, 46T12, 60J65, 60J25

Library of Congress Control Number: 2006924566

ISBN-10 3-540-29020-6 Springer Berlin Heidelberg New York
ISBN-13 978-3-540-29020-9 Springer Berlin Heidelberg New York

Springer is a part of Springer Science+Business Media
springer.com
© Springer-Verlag Berlin Heidelberg 2006
Printed in Germany

Cover design: Erich Kirchner, Heidelberg
Typesetting by the author and SPI Publisher Services using a Springer LaTeX macro package

Printed on acid-free paper   11500582   41/sz - 5 4 3 2 1 0

Preface

This volume is a revised and extended version of the lecture notes concerning a one-year course on infinite dimensional analysis delivered at Scuola Normale Superiore in recent years, see [6]. The lectures were designed for an audience having basic knowledge of functional analysis and measure theory but not familiar with probability theory.

The main aim was to give an introduction to the analysis in a separable Hilbert space H of infinite dimensions. It is well known that there is no natural analogue of the Lebesgue measure on an infinite dimensional Hilbert space. A natural substitute is provided by *Gaussian measures* which are introduced in Chapter 1. They are first defined on a finite dimensional space and then, through an infinite product of measures, on the infinite dimensional Hilbert space H. Moreover, given a Gaussian measure, the useful concepts of *Cameron–Martin space* and *white noise* mapping are presented.

Chapter 2 is devoted to the *Cameron–Martin* formula. This is a basic tool in discussing absolute continuity and singularity of a Gaussian measure and its translates.

In Chapter 3 we introduce in a simple way, using the white noise mapping, *Brownian motion* and the *Wiener integral*. Using these concepts, in Chapter 4 we consider dynamical systems (governed by ordinary differential equations) perturbed by white noise and introduce the corresponding transition semigroups. It is the first example of a Markov semigroup we meet in this course. General Markov semigroups are presented in Chapter 5 where their asymptotic behaviour as $t \to \infty$ is studied introducing concepts such as: *invariant measures, ergodicity, mixing*. Chapter 6 is devoted to the *Prokhorov* theorem about weak convergence of measures, and Chapter 7 to existence and uniqueness of invariant measures including theorems due to *Krylov–Bogoliubov* and

Khas'minskii. These results are illustrated in Chapter 8, through a discussion of the infinite dimensional *Heat* equation and the *Ornstein–Uhlenbeck* semigroup.

In Chapter 9, we consider the space $L^2(H, \mu)$ where μ is a nondegenerate Gaussian measure on H, proving the *Itô–Wiener* decomposition in terms of generalized *Hermite polynomials*. Then in Chapter 10 we define the Sobolev spaces $W^{1,2}(H, \mu)$ and $W^{2,2}(H, \mu)$; we prove that the embedding $L^2(H, \mu) \subset W^{1,2}(H, \mu)$ is compact and that the *Poincaré* and *log-Sobolev* inequalities hold. Finally, Chapter 11 is devoted to an introduction to gradient systems. Here we start from a nondegenerate Gaussian measure μ on H and define a probability measure of the form

$$\nu(dx) = Z^{-1} e^{-U(x)} \nu(dx), \quad x \in H,$$

where U is a *potential* and Z a normalization constant. A typical example of such a measure is provided by the *Gibbs* measures, see e.g. [25].

We show that there exists a transition semigroup P_t for which the measure ν is invariant and extend to P_t several properties of the Ornstein–Uhlenbeck semigroup.

An appendix about linear operators, including closed and closable operators, cores, strongly continuous semigroups, Hille–Yosida and Lumer–Phillips theorems, has been added in order to make the book as self-contained as possible.

The text [6] has been completely revised, in particular some proofs have been simplified and several misprints and mistakes have been corrected. Moreover, several details have been added as well as some new material: Dynamical systems with dissipative nonlinearities in Chapter 7 and the asymptotic behaviour (including ergodicity, mixing and spectral gap) for gradient systems.

Several concepts and results contained in the book are taken from the monographs [9], [10] and [11].

Pisa, 1 January 2006 Giuseppe Da Prato

Contents

Gaussian measures in Hilbert spaces

1.1 Notations and preliminaries

We assume that the reader is familiar with the basic notions of probability theory and functional analysis; some of them will be recalled when needed. One can look for instance at the books by P. Billingsley [1], L. Breiman [2] and M. Métivier [21].

Let $(\Omega, \mathscr{F}, \mathbb{P})$ be a probability space. For any $A \in \mathscr{F}$ we shall denote by A^c the complement of A and by $\mathbf{1}_A$ the function

$$\mathbf{1}_A(\omega) = \begin{cases} 1 & \text{if } \omega \in A \\ 0 & \text{if } \omega \in A^c. \end{cases}$$

Moreover, given any complete metric space E, we shall denote by $\mathscr{B}(E)$ the σ-algebra generated by all closed (or equivalently open) subsets of E.

By a *random variable* in $(\Omega, \mathscr{F}, \mathbb{P})$ with values in E we mean a mapping $X \colon \Omega \to E$ such that

$$I \in \mathscr{B}(E) \Rightarrow X^{-1}(I) \in \mathscr{F}.$$

The *law* of X is the probability measure $X_\# \mathbb{P}$ on $(E, \mathscr{B}(E))$ defined as

$$X_\# \mathbb{P}(I) = \mathbb{P}(X^{-1}(I)) = \mathbb{P}(X \in I), \quad I \in \mathscr{B}(E).$$

We recall the basic change of variables formula,

Proposition 1.1 *Let X be a random variable in $(\Omega, \mathscr{F}, \mathbb{P})$ with values in E. Let moreover $\varphi \colon E \to \mathbb{R}$ be a bounded Borel mapping. Then we have*

$$\int_\Omega \varphi(X(\omega))\mathbb{P}(d\omega) = \int_E \varphi(x) X_\# \mathbb{P}(dx). \tag{1.1}$$

Proof. It is enough to prove (1.1) when $\varphi = 1_I$ and $I \in \mathscr{B}(E)$. In this case we have
$$\varphi(X(\omega)) = 1_{X^{-1}(I)}(\omega), \quad \omega \in \Omega.$$

So,
$$\int_\Omega \varphi(X(\omega))\mathbb{P}(d\omega) = \mathbb{P}(X^{-1}(I)) = X_\# \mathbb{P}(I) = \int_E \varphi(x) X_\# \mathbb{P}(dx).$$

$\square$

This chapter is devoted to the concept of *Gaussian measure* on $(H, \mathscr{B}(H))$, where H is a real separable Hilbert space with inner product $\langle \cdot, \cdot \rangle$ and norm $|\cdot|$. We denote by $L(H)$ the Banach algebra of all continuous linear operators from H into H, by $L^+(H)$ the set of all $T \in L(H)$ which are symmetric ($\langle Tx, y \rangle = \langle x, Ty \rangle, x, y \in H$) and positive ($\langle Tx, x \rangle \geq 0, x \in H$).

Finally, we denote by $L_1^+(H)$ the set of all operators $Q \in L^+(H)$ of *trace class* that is such that $\operatorname{Tr} Q := \sum_{k=1}^\infty \langle Qe_k, e_k \rangle < +\infty$ for one (and consequently for all) complete orthonormal system (e_k) in H. If Q is of trace class then it is compact and $\operatorname{Tr} Q$ is the sum of its eigenvalues repeated according to their multiplicity, see e.g. [15].

A Gaussian measure will be defined in section 1.2 when H is one-dimensional and in section 1.3 when it is finite dimensional. Section 1.4 is devoted to some general properties of Borel measures in Hilbert spaces and section 1.5 to the definition of Gaussian measure in a infinite dimensional Hilbert space. Finally, section 1.6 is devoted to the concept of *Gaussian random variable* and section 1.7 to the definition of *Cameron–Martin space* and *white noise* mapping.

1.2 One-dimensional Hilbert spaces

For any couple of real numbers (a, λ) with $a \in \mathbb{R}$ and $\lambda \geq 0$ we define a probability measure $N_{a,\lambda}$ in $(\mathbb{R}, \mathscr{B}(\mathbb{R}))$ as follows. If $\lambda = 0$ we set
$$N_{a,0} = \delta_a,$$

where δ_a is the Dirac measure at a,
$$\delta_a(B) = \begin{cases} 1 & \text{if } a \in B, \\ 0 & \text{if } a \notin B, \end{cases} \quad B \in \mathscr{B}(\mathbb{R}).$$

If $\lambda > 0$ we set

$$N_{a,\lambda}(B) = \frac{1}{\sqrt{2\pi\lambda}} \int_B e^{-\frac{(x-a)^2}{2\lambda}} \, dx, \quad B \in \mathscr{B}(\mathbb{R}).$$

$N_{a,\lambda}$ is a probability measure in $(\mathbb{R}, \mathscr{B}(\mathbb{R}))$ since

$$N_{a,\lambda}(\mathbb{R}) = \frac{1}{\sqrt{2\pi\lambda}} \int_{-\infty}^{+\infty} e^{-\frac{(x-a)^2}{2\lambda}} \, dx = \frac{1}{\sqrt{2\pi}} \int_{-\infty}^{+\infty} e^{-\frac{x^2}{2}} \, dx = 1.$$

If $\lambda > 0$, $N_{a,\lambda}$ is absolutely continuous with respect to the Lebesgue measure. In this case we shall write

$$N_{a,\lambda}(dx) = \frac{1}{\sqrt{2\pi\lambda}} e^{-\frac{(x-a)^2}{2\lambda}} \, dx.$$

When $a = 0$ we shall write N_λ instead of $N_{a,\lambda}$ for short.

In the following proposition we list some basic properties of the probability $N_{a,\lambda}$; the simple proofs are left to the reader as an exercise.

Proposition 1.2 *Let $a \in \mathbb{R}$ and $\lambda > 0$. Then we have*

$$\int_{\mathbb{R}} x N_{a,\lambda}(dx) = a,$$

$$\int_{\mathbb{R}} (x-a)^2 N_{a,\lambda}(dx) = \lambda,$$

$$\widehat{N_{a,\lambda}}(h) := \int_{\mathbb{R}} e^{ihx} N_{a,\lambda}(dx) = e^{iah - \frac{1}{2}\lambda h^2}, \quad h \in \mathbb{R}.$$

a is called the *mean*, λ the *variance*, and $\widehat{N_{a,\lambda}}$ the *Fourier transform* (or the *characteristic function*) of $N_{a,\lambda}$.

1.3 Finite dimensional Hilbert spaces

We assume here that H has finite dimension equal to $d \in \mathbb{N}$. Before defining Gaussian measures on $(H, \mathscr{B}(H))$ we recall some concepts on product probabilities.

1.3.1 Product probabilities

Let $d \in \mathbb{N}$ and $(\Omega_i, \mathscr{F}_i, \mathbb{P}_i)$, $i = 1, \ldots, d$, be probability spaces. Denote by Ω the product space $\Omega := \Omega_1 \times \Omega_2 \times \cdots \times \Omega_d$. A subset R of Ω is said to be a *measurable rectangle* if it is of the form

$$R = B_1 \times B_2 \times \cdots \times B_d, \quad B_i \in \mathscr{F}_i, \ i = 1, \ldots, d.$$

The σ-algebra $\mathscr{F}$ generated by the set of all measurable rectangles is called the *product σ-algebra* of $\mathscr{F}_i$, $i = 1, \ldots, d$. For any measurable rectangle $R = B_1 \times B_2 \times \ldots \times B_d$, we define

$$\mathbb{P}(R) = \mathbb{P}_1(B_1) \times \mathbb{P}_2(B_2) \times \cdots \times \mathbb{P}_d(B_d).$$

Then $\mathbb{P}$ can be uniquely extended to a probability measure on $(\Omega, \mathscr{F})$, denoted by $\mathbb{P}_1 \times \cdots \times \mathbb{P}_d$, see e.g. [18].

1.3.2 Definition of Gaussian measures

We are going to define a Gaussian measure $N_{a,Q}$ for any $a \in H$ and any $Q \in L^+(H)$.

Let $Q \in L^+(H)$ and let $(e_1, \ldots, e_d)$ be an orthonormal basis on H such that $Qe_k = \lambda_k e_k$, $k = 1, \ldots, d$, for some $\lambda_k \geq 0$. We set

$$x_k = \langle x, e_k \rangle, \quad x \in H, \ k = 1, \ldots, d,$$

and we identify H with $\mathbb{R}^d$ through the isomorphism γ,

$$\gamma : H \to \mathbb{R}^d, \quad x \mapsto \gamma(x) = (x_1, \ldots, x_d).$$

Now we define a probability measure $N_{a,Q}$ on $(\mathbb{R}^d, \mathscr{B}(\mathbb{R}^d))$ by setting

$$N_{a,Q} = \mathop{\bigtimes}_{k=1}^{d} N_{a_k, \lambda_k}.$$

When $a = 0$ we shall write N_Q instead of $N_{a,Q}$ for short.

The proof of the following proposition is easy; it is left to the reader.

Proposition 1.3 *Let $a \in H$, $Q \in L^+(H)$ and $\mu = N_{a,Q}$. Then we have*

$$\int_H x N_{a,Q}(dx) = a,$$

$$\int_H \langle y, x - a \rangle \langle z, x - a \rangle N_{a,Q}(dx) = \langle Qy, z \rangle, \quad y, z \in H.$$

Moreover the Fourier tranform of $N_{a,Q}$ is given by

$$\widehat{N_{a,Q}}(h) := \int_H e^{i\langle h, x \rangle} N_{a,Q}(dx) = e^{i\langle a, h \rangle - \frac{1}{2}\langle Qh, h \rangle}, \quad h \in H.$$

Finally, if the determinant of Q is positive, $N_{a,Q}$ is absolutely continuous with respect to the Lebesgue measure in $\mathbb{R}^d$ and we have

$$N_{a,Q}(dx) = \frac{1}{\sqrt{(2\pi)^d \det Q}} e^{-\frac{1}{2}\langle Q^{-1}(x-a), x-a \rangle} dx.$$

a is called the *mean* and Q the *covariance operator* of $N_{a,Q}$.

The following result is a consequence of the uniqueness of the Fourier transform, see for instance [21, page 154].

Proposition 1.4 *Let H be a finite dimensional Hilbert space. Let $a \in H$, $Q \in L^+(H)$ and let μ be a probability measure on $(H, \mathcal{B}(H))$ such that*

$$\int_H e^{i\langle h, x\rangle} \mu(dx) = e^{i\langle a, h\rangle - \frac{1}{2}\langle Qh, h\rangle}, \quad h \in H.$$

Then $\mu = N_{a,Q}$.

1.4 Measures in Hilbert spaces

In this section H represents an infinite dimensional separable Hilbert space and (e_k) a complete orthonormal system in H. For any $n \in \mathbb{N}$ we consider the projection mapping $P_n \colon H \to P_n(H)$ defined as

$$P_n x = \sum_{k=1}^n \langle x, e_k\rangle e_k, \quad x \in H. \tag{1.2}$$

Obviously we have $\lim_{n\to\infty} P_n x = x$ for all $x \in H$.

Proposition 1.5 *Let $\mu, \nu \in M(H)$ be such that*

$$\int_H \varphi(x)\mu(dx) = \int_H \varphi(x)\nu(dx), \tag{1.3}$$

for all continuous and bounded functions $\varphi \colon H \to \mathbb{R}$ in H. Then $\mu = \nu$.

Proof. Let $C \subset H$ be closed and let (φ_n) be a sequence of continuous and bounded functions in H such that

$$\begin{cases} \varphi_n(x) \to \mathbf{1}_C(x), & x \in H, \\ \sup_{x \in H} |\varphi_n(x)| \le 1, \end{cases} \tag{1.4}$$

where $\mathbf{1}_C$ is the characteristic function of C. A sequence $(\varphi_n) \subset C_b(H)$ such that (1.4) holds is provided by,

$$\varphi_n(x) = \begin{cases} 1 & \text{if } x \in C, \\ 1 - n\, d(x, C) & \text{if } d(x, C) \le \frac{1}{n} \\ 0 & \text{if } d(x, C) \ge \frac{1}{n}. \end{cases}$$

Now, by the dominated convergence theorem it follows that

$$\int_H \varphi_n d\mu = \int_H \varphi_n d\nu \to \mu(C) = \nu(C).$$

Since closed sets generate the Borel σ-algebra of H this implies that $\mu = \nu$. $\square$

Proposition 1.6 *Let μ and ν be probability measures on $(H, \mathscr{B}(H))$. If $(P_n)_{\#}\mu = (P_n)_{\#}\nu$ for any $n \in \mathbb{N}$ we have $\mu = \nu$.*

Proof. Let $\varphi \colon H \to \mathbb{R}$ be continuous and bounded. Then, by the dominated convergence theorem, we have

$$\int_H \varphi(x)\mu(dx) = \lim_{n\to\infty} \int_H \varphi(P_n x)\mu(dx).$$

Therefore, by the change of variables formula (1.1), it follows that

$$\int_H \varphi(x)\mu(dx) = \lim_{n\to\infty} \int_H \varphi(P_n x)\mu(dx) = \lim_{n\to\infty} \int_{P_n(H)} \varphi(\xi)(P_n)_{\#}\mu(d\xi)$$

$$= \lim_{n\to\infty} \int_{P_n(H)} \varphi(\xi)(P_n)_{\#}\nu(d\xi) = \lim_{n\to\infty} \int_H \varphi(P_n x)\nu(dx) = \int_H \varphi(x)\nu(dx).$$

So, by the arbitrariness of φ, we have $\mu = \nu$ thanks to Proposition 1.5. $\square$

Let us consider now the Fourier transform of μ,

$$\widehat{\mu}(h) \colon = \int_H e^{i\langle x,h\rangle}\mu(dx), \quad h \in H.$$

Proposition 1.7 *Let μ and ν be probability measures on $(H, \mathscr{B}(H))$. Then if $\widehat{\mu}(h) = \widehat{\nu}(h)$ for all $h \in H$, we have $\mu = \nu$.*

Proof. For any $n \in \mathbb{N}$, we have by (1.1),

$$\widehat{\mu}(P_n h) = \int_H e^{i\langle x,P_n h\rangle}\mu(dx) = \int_{P_n(H)} e^{i\langle P_n\xi,P_n h\rangle}(P_n)_{\#}\mu(d\xi)$$

and

$$\widehat{\nu}(P_n h) = \int_H e^{i\langle x,P_n h\rangle}\nu(dx) = \int_{P_n(H)} e^{i\langle P_n\xi,P_n h\rangle}(P_n)_{\#}\nu(d\xi).$$

Since $\widehat{\mu}(P_n h) = \widehat{\nu}(P_n h)$ by assumption, the measures $(P_n)_{\#}\mu$ and $(P_n)_{\#}\nu$ have the same Fourier transforms and so they coincide from Proposition 1.4. The conclusion follows now from Proposition 1.6. $\square$

Let us now fix a probability measure μ on $(H, \mathscr{B}(H))$. We are going to define the *mean* and the *covariance* of μ.

Assume that

$$\int_H |x|\mu(dx) < +\infty.$$

Then for any $h \in H$ the linear functional $F\colon H \to \mathbb{R}$ defined as

$$F(h) = \int_H \langle x, h \rangle \mu(dx), \quad h \in H,$$

is continuous since

$$|F(h)| \le \int_H |x|\mu(dx)\,|h|, \quad h \in H.$$

By the Riesz representation theorem there exists $m \in H$ such that

$$\langle m, h \rangle = \int_H \langle x, h \rangle \mu(dx), \quad h \in H.$$

m is called the *mean* of μ. We shall write

$$m = \int_H x\mu(dx).$$

Assume now that

$$\int_H |x|^2 \mu(dx) < +\infty.$$

Thus we can consider the bilinear form $G\colon H \times H \to \mathbb{R}$ defined as

$$G(h, k) = \int_H \langle h, x - m \rangle \langle k, x - m \rangle \mu(dx), \quad h, k \in H.$$

G is continuous since

$$|G(h, k)| \le \int_H |x - m|^2 \mu(dx)\,|h|\,|k|, \quad h, k \in H.$$

Therefore, again by the Riesz theorem, there is a unique linear bounded operator $Q \in L(H)$ such that

$$\langle Qh, k \rangle = \int_H \langle h, x - m \rangle \langle k, x - m \rangle \mu(dx), \quad h, k \in H.$$

Q is called the *covariance* of μ.

Proposition 1.8 *Let μ be a probability measure on $(H, \mathscr{B}(H))$ with mean m and covariance Q. Then $Q \in L_1^+(H)$, i.e. Q is symmetric, positive and of trace class.*

Proof. Symmetry and positivity of Q are clear. To prove that Q is of trace class, write

$$\langle Qe_k, e_k \rangle = \int_H |\langle x - m, e_k \rangle|^2 \mu(dx), \quad k \in \mathbb{N}.$$

By the monotone convergence theorem and the Parseval identity, we find that

$$\operatorname{Tr} Q = \sum_{k=1}^{\infty} \int_H |\langle x - m, e_k \rangle|^2 \mu(dx) = \int_H |x - m|^2 \mu(dx) < \infty.$$

$\square$

1.5 Gaussian measures

Let $a \in H$ and $Q \in L_1^+(H)$. A *Gaussian measure* $\mu := N_{a,Q}$ on $(H, \mathscr{B}(H))$ is a measure μ having mean a, covariance operator Q and Fourier transform given by

$$\widehat{N_{a,Q}}(h) = \exp\left\{ i\langle a, h \rangle - \frac{1}{2} \langle Qh, h \rangle \right\}, \quad h \in H. \tag{1.5}$$

The Gaussian measure $N_{a,Q}$ is said to be *nondegenerate* if Ker $(Q) = \{x \in H : Qx = 0\} = \{0\}$.

We are going to show that for arbitrary $a \in H$ and $Q \in L_1^+(H)$ there exists a unique Gaussian measure $\mu = N_{a,Q}$ in $(H, \mathscr{B}(H))$.

First notice that, since $Q \in L_1^+(H)$, there exists a complete orthonormal basis (e_k) on H and a sequence of non-negative numbers (λ_k) such that

$$Qe_k = \lambda_k e_k, \quad k \in \mathbb{N}.$$

For any $x \in H$ we set $x_k = \langle x, e_k \rangle$, $k \in \mathbb{N}$.

Let us consider the natural isomorphism γ between H and the Hilbert space ℓ^2 of all sequences (x_k) of real numbers such that

$$\sum_{k=1}^{\infty} |x_k|^2 < \infty,$$

defined by

$$H \to \ell^2, \quad x \mapsto \gamma(x) = (x_k).$$

In this section we shall identify H with ℓ^2. Then we shall consider the product measure

$$\mu := \underset{k=1}{\overset{\infty}{\times}} N_{a_k, \lambda_k}.$$

Though μ is defined on the space $\mathbb{R}^\infty := \times_{k=1}^{\infty} \mathbb{R}$ (instead on ℓ^2), we shall show that it is concentrated on ℓ^2. Finally, we shall prove that μ is a Gaussian measure $N_{a,Q}$.

We need some results on countable product of measures which we recall in the next subsection.

1.5.1 Some results on countable product of measures

Let (μ_k) be a sequence of probability measures on $(\mathbb{R}, \mathscr{B}(\mathbb{R}))$. We want to define a product measure on the space $\mathbb{R}^\infty = \times_{k=1}^{\infty} \mathbb{R}$, consisting of all sequences $x = (x_k)$ of real numbers. We shall endow $\mathbb{R}^\infty$ with the distance

$$d(x,y) = \sum_{n=1}^{\infty} 2^{-n} \frac{\max\{|x_k - y_k| : 1 \le k \le n\}}{1 + \max\{|x_k - y_k| : 1 \le k \le n\}}.$$

It is easy to see that $\mathbb{R}^\infty$, with this distance, is a complete metric space and that the topology corresponding to its metric is precisely the product topology.

We are going to define $\mu = \times_{k=1}^{\infty} \mu_k$ on the family $\mathscr{C}$ of all *cylindrical subsets* $I_{n,A}$ of $\mathbb{R}^\infty$, where $n \in \mathbb{N}$ and $A \in \mathscr{B}(\mathbb{R}^n)$,

$$I_{n,A} = \{x = (x_k) \in \mathbb{R}^\infty : (x_1, \dots, x_n) \in A\}.$$

Notice that

$$I_{n,A} = I_{n+k, A \times X_{n+1} \times \cdots \times X_{n+k}}, \quad k, n \in \mathbb{N}.$$

Using this identity one can easily see that $\mathscr{C}$ is an algebra. In fact, if $I_{n,A}$ and $I_{m,B}$ are two cylindrical sets we have

$$I_{n,A} \cup I_{m,B} = I_{m+n, A \times X_{n+1} \times \cdots \times X_{n+m}} \cup I_{m+n, B \times X_{m+1} \times \cdots \times X_{m+n}}$$

(1.6)

$$= I_{m+n, A \times X_{n+1} \times \cdots \times X_{n+m} \cup B \times X_{m+1} \times \cdots \times X_{m+n}},$$

and $I_{n,A}^c = I_{n,A^c}$.

Moreover, the σ-algebra generated by $\mathscr{C}$ coincides with $\mathscr{B}(\mathbb{R}^\infty)$ since any ball (with respect to the metric of $\mathbb{R}^\infty$) is a countable intersection of cylindrical sets.

Finally, we define the product measure

$$\mu(I_{n,A}) = (\mu_1 \times \cdots \times \mu_n)(A), \quad I_{n,A} \in \mathscr{C}.$$

Using (1.6) we see that μ is additive. We want now to show that μ is σ-additive on $\mathscr{C}$. This will imply by the Caratheodory extension theorem, see e.g. [18], that μ can be uniquely extended to a probability measure on the product σ-algebra $\mathscr{B}(\mathbb{R}^\infty)$.

Theorem 1.9 μ *is σ-additive on $\mathscr{C}$ and it possesses a unique extension to a probability measure on $(\mathbb{R}^\infty, \mathscr{B}(\mathbb{R}^\infty))$.*

Proof. To prove σ-additivity of μ it is enough to show continuity of μ at $\varnothing$. This is equivalent to prove that if (E_j) is a decreasing sequence on $\mathscr{C}$ such that

$$\mu(E_j) \geq \varepsilon, \quad j \in \mathbb{N},$$

for some $\varepsilon > 0$, we have

$$\bigcap_{j=1}^{\infty} E_j \neq \varnothing.$$

To prove this fact, let us consider the following sections of E_j,

$$E_j(\alpha) = \{x \in \mathbb{R}_1^\infty : (\alpha, x) \in E_j\}, \quad \alpha \in \mathbb{R},$$

where we have used the notation $\mathbb{R}_n^\infty = \bigtimes_{k=n+1}^{\infty} \mathbb{R}$, $n \in \mathbb{N}$. Set

$$F_j^{(1)} = \left\{\alpha \in \mathbb{R} : \mu^{(1)}(E_j(\alpha)) \geq \frac{\varepsilon}{2}\right\}, \quad j \in \mathbb{N},$$

where $\mu^{(n)} = \bigtimes_{k=n+1}^{\infty} \mu_k$, $n \in \mathbb{N}$. Then by the Fubini theorem we have

$$\mu(E_j) = \int_{\mathbb{R}} \mu^{(1)}(E_j(\alpha))\mu_1(d\alpha)$$

$$= \int_{F_j^{(1)}} \mu^{(1)}(E_j(\alpha))\mu_1(d\alpha) + \int_{[F_j^{(1)}]^c} \mu^{(1)}(E_j(\alpha))\mu_1(d\alpha)$$

$$\leq \mu_1(F_j^{(1)}) + \frac{\varepsilon}{2}.$$

Therefore

$$\mu_1(F_j^{(1)}) \geq \frac{\varepsilon}{2}.$$

Since μ_1 is a probability measure on $(\mathbb{R}, \mathscr{B}(\mathbb{R}))$, it is continuous at $\varnothing$. Therefore, since the sequence $(F_j^{(1)})$ is decreasing, there exists $\overline{\alpha_1} \in \mathbb{R}$ such that

$$\mu^1(E_j(\overline{\alpha_1})) \geq \frac{\varepsilon}{2}, \quad j \in \mathbb{N},$$

and consequently we have

$$E_j(\overline{\alpha_1}) \neq \varnothing. \tag{1.7}$$

Now set

$$E_j(\overline{\alpha_1}, \alpha_2) = \{x_2 \in \mathbb{R}_2^\infty : (\overline{\alpha_1}, \alpha_2, x) \in E_j\}, \quad j \in \mathbb{N}, \ \alpha_2 \in \mathbb{R},$$

and

$$F_j^{(2)} = \left\{\alpha_2 \in \mathbb{R} : \mu^{(2)}(E_j(\alpha)) \geq \frac{\varepsilon}{2}\right\}, \quad j \in \mathbb{N}.$$

Then, again by the Fubini theorem, we have

$$\mu^1(E_j(\overline{\alpha_1})) = \int_{\mathbb{R}} \mu^{(2)}(E_j(\overline{\alpha_1}, \alpha_2))\mu_2(d\alpha_2)$$

$$= \int_{F_j^{(2)}} \mu^{(2)}(E_j(\overline{\alpha_1}, \alpha_2))\mu_2(d\alpha_2)$$

$$+ \int_{[F_j^{(2)}]^c} \mu^{(2)}(E_j(\overline{\alpha_1}, \alpha_2))\mu_2(d\alpha_2)$$

$$\leq \mu_2(F_j^{(2)}) + \frac{\varepsilon}{4}.$$

Therefore

$$\mu_2(F_j^{(2)}) \geq \frac{\varepsilon}{4}.$$

Since $(F_j^{(2)})$ is decreasing, there exists $\overline{\alpha_2} \in \mathbb{R}$ such that

$$\mu^2(E_j(\overline{\alpha_1}, \overline{\alpha_2})) \geq \frac{\varepsilon}{4}, \quad j \in \mathbb{N},$$

and consequently we have

$$E_j(\overline{\alpha_1}, \overline{\alpha_2}) \neq \varnothing. \tag{1.8}$$

Arguing in a similar way as before we see that there exists a sequence $(\overline{\alpha_k}) \in \mathbb{R}^\infty$ such that

$$E_j(\overline{\alpha_1}, \ldots, \overline{\alpha_n}) \neq \varnothing, \tag{1.9}$$

where

$$E_j(\alpha_1, \ldots, \alpha_n) = \{x \in \mathbb{R}_n^\infty : (\alpha_1, \ldots, \alpha_n, x) \in E_j\}, \quad n \in \mathbb{N}.$$

This implies, as easily seen,

$$(\alpha_n) \in \bigcap_{j=1}^{\infty} E_j.$$

Therefore $\bigcap_{j=1}^{\infty} E_j$ is not empty as required. Thus we have proved that μ is σ-additive on $\mathscr{C}$ and consequently on $\mathscr{B}(\mathbb{R}^{\infty})$. $\square$

1.5.2 Definition of Gaussian measures

We are going to show that

$$\mu = \mathop{\mathsf{X}}_{k=1}^{\infty} N_{a_k, \lambda_k} \tag{1.10}$$

is a Gaussian measure on $H = \ell^2$ with mean a and covariance Q.
We first show that μ is concentrated in ℓ^2.

Exercise 1.10 Prove that ℓ^2 is a Borel subset of $\mathbb{R}^{\infty}$.

Proposition 1.11 We have $\mu(\ell^2) = 1$.

Proof. From the monotone convergence theorem we have in fact

$$\int_{\mathbb{R}^{\infty}} \sum_{k=1}^{\infty} x_k^2 \mu(dx) = \sum_{k=1}^{\infty} \int_{\mathbb{R}} x_k^2 N_{a_k, \lambda_k}(dx_k) = \sum_{k=1}^{\infty} (\lambda_k + a_k^2). \tag{1.11}$$

Therefore

$$\mu\left(\left\{x \in \mathbb{R}^{\infty} : |x|_{\ell^2}^2 < \infty\right\}\right) = 1.$$

$\square$

Theorem 1.12 There exists a unique probability measure μ on $(H, \mathscr{B}(H))$ with mean a, covariance Q and Fourier transform

$$\widehat{\mu}(h) = \exp\left\{i\langle a, h\rangle - \frac{1}{2}\langle Qh, h\rangle\right\}, \quad h \in H. \tag{1.12}$$

μ will be denoted by $N_{a,Q}$. If $a = 0$ we shall write N_Q instead of $N_{0,Q}$.

Proof. We check that the restriction to ℓ^2 of the product measure μ, defined by (1.10), fulfills the required properties.
First notice that by (1.11) we have

$$\int_{H} |x|^2 \mu(dx) = \operatorname{Tr} Q + |a|^2. \tag{1.13}$$

From now on we assume, for the sake of simplicity, that Ker $(Q) = \{0\}$ and (this is not a restriction) that

$$\lambda_1 \geq \lambda_2 \geq \cdots \geq \lambda_n \geq \cdots$$

Let (P_n) be the sequence of projectors defined by (1.2) and let $h \in H$. Since $|\langle x, h \rangle| \leq |x|\,|h|$ and $\int_H |x|\mu(dx)$ is finite by (1.13), we have, by the dominated convergence theorem,

$$\int_H \langle x, h \rangle \mu(dx) = \lim_{n \to \infty} \int_H \langle P_n x, h \rangle \mu(dx).$$

But

$$\int_H \langle P_n x, h \rangle \mu(dx) = \sum_{k=1}^{n} \int_H x_k h_k \mu(dx)$$

$$= \sum_{k=1}^{n} h_k \int_{\mathbb{R}} x_k N_{a_k, \lambda_k}(dx_k) = \sum_{k=1}^{n} h_k a_k = \langle P_n a, h \rangle \to \langle a, h \rangle,$$

as $n \to \infty$. Thus the mean of μ is a.

In a similar way, to determine the covariance of μ we fix $y, z \in H$ and write

$$\int_H \langle x - a, y \rangle \langle x - a, z \rangle \mu(dx) = \lim_{n \to \infty} \int_H \langle P_n(x - a), y \rangle \langle P_n(x - a), z \rangle \mu(dx).$$

Moreover

$$\int_H \langle P_n(x-a), y \rangle \langle P_n(x-a), z \rangle \mu(dx)$$

$$= \sum_{k=1}^{n} \int_H (x_k - a_k)^2 y_k z_k \mu(dx)$$

$$= \sum_{k=1}^{n} y_k z_k \int_{\mathbb{R}} (x_k - a_k)^2 N_{a_k, \lambda_k}(dx_k)$$

$$= \sum_{k=1}^{n} y_k z_k \lambda_k = \langle P_n Q y, z \rangle \to \langle Q y, z \rangle,$$

as $n \to \infty$. So, the covariance of μ is equal to Q.

Finally, for any $h \in H$ we have

$$\int_H e^{i\langle x, h \rangle} \mu(dx) = \lim_{n \to \infty} \int_H e^{i\langle P_n x, h \rangle} \mu(dx)$$

$$= \lim_{n \to \infty} \prod_{k=1}^{n} \int_{\mathbb{R}} e^{i x_k h_k} N_{a_k, \lambda_k}(dx_k)$$

$$= \lim_{n \to \infty} \prod_{k=1}^{n} e^{i a_k h_k - \frac{1}{2} \lambda_k h_k^2} = \lim_{n \to \infty} e^{i\langle P_n a, h \rangle} e^{-\frac{1}{2} \langle P_n Q h, h \rangle}$$

$$= e^{i\langle a, h \rangle} e^{-\frac{1}{2} \langle Q h, h \rangle}.$$

So, the Fourier transform of μ is given by (1.12). The proof is complete. $\square$

We conclude this section by computing some Gaussian integrals which will be useful later. We still assume that $\lambda_1 \geq \lambda_2 \geq \cdots \geq \lambda_n \geq \cdots$.

To formulate the next result, notice that for any $\varepsilon < \frac{1}{\lambda_1}$, the linear operator $1 - \varepsilon Q$ is invertible and $(1 - \varepsilon Q)^{-1}$ is bounded. We have in fact, as easily checked,

$$(1 - \varepsilon Q)^{-1}x = \sum_{k=1}^{\infty} \frac{1}{1 - \varepsilon \lambda_k} \langle x, e_k \rangle e_k, \quad x \in H.$$

In this case we can define the *determinant* of $(1 - \varepsilon Q)$ by setting

$$\det(1 - \varepsilon Q) := \lim_{n \to \infty} \prod_{k=1}^{n} (1 - \varepsilon \lambda_k) := \prod_{k=1}^{\infty} (1 - \varepsilon \lambda_k).$$

It is easy to see that, in view of the assumption $\sum_{k=1}^{\infty} \lambda_k < +\infty$, the infinite product above is finite and positive.

Proposition 1.13 *Let $\varepsilon \in \mathbb{R}$. Then we have*

$$\int_H e^{\frac{\varepsilon}{2}|x|^2} \mu(dx) = \begin{cases} [\det(1 - \varepsilon Q)]^{-1/2} e^{-\frac{\varepsilon}{2}\langle (1-\varepsilon Q)^{-1}a,a\rangle}, & \text{if } \varepsilon < \frac{1}{\lambda_1}, \\ +\infty, & \text{otherwise.} \end{cases}$$
(1.14)

Proof. For any $n \in \mathbb{N}$ we have

$$\int_H e^{\frac{\varepsilon}{2}|P_n x|^2} \mu(dx) = \prod_{k=1}^{n} \int_{\mathbb{R}} e^{\frac{\varepsilon}{2}x_k^2} N_{a_k, \lambda_k}(dx_k).$$

Since $|P_n x|^2 \uparrow |x|^2$ as $n \to \infty$ and, by an elementary computation,

$$\int_{\mathbb{R}} e^{\frac{\varepsilon}{2}x_k^2} N_{a_k, \lambda_k}(dx_k) = \frac{1}{\sqrt{1 - \varepsilon \lambda_k}} e^{-\frac{\varepsilon}{2} \frac{a_k^2}{1-\varepsilon \lambda_k}},$$

the conclusion follows from the monotone convergence theorem. $\square$

Exercise 1.14 Compute the integral

$$J_m = \int_H |x|^{2m} \mu(dx), \quad m \in \mathbb{N}.$$

Hint. Notice that $J_m = 2^m F^{(m)}(0)$, where

$$F(\varepsilon) = \int_H e^{\frac{\varepsilon}{2}|x|^2} \mu(dx), \quad \varepsilon > 0.$$

Proposition 1.15 *We have*

$$\int_H e^{\langle h,x\rangle}\mu(dx) = e^{\langle a,h\rangle}e^{\frac{1}{2}\langle Qh,h\rangle}, \quad h \in H. \tag{1.15}$$

Proof. For any $\varepsilon > 0$ we have

$$e^{\langle h,x\rangle} \le e^{|x|\,|h|} \le e^{\varepsilon|x|^2}\,e^{\frac{1}{\varepsilon}|h|^2}.$$

Choosing $\varepsilon < \frac{1}{\lambda_1}$, we have, by the dominated convergence theorem,

$$\int_H e^{\langle h,x\rangle}\mu(dx) = \lim_{n\to\infty}\int_H e^{\langle h,P_n x\rangle}\mu(dx) = \lim_{n\to\infty}\int_H e^{\langle h,P_n x\rangle}\underset{j=1}{\overset{n}{\times}} N_{a_j\lambda_j}(dx)$$

$$= \lim_{n\to\infty} e^{\frac{1}{2}\langle P_n Qh,h\rangle} = e^{\frac{1}{2}\langle Qh,h\rangle}.$$

$\square$

1.6 Gaussian random variables

Let $(\Omega, \mathscr{F}, \mathbb{P})$ be a probability space, K a separable Hilbert space and X a random variable in $(\Omega, \mathscr{F}, \mathbb{P})$ with values in K. If the law of X is Gaussian we say that X is a *Gaussian random variable* with values in K.

For any $p \ge 1$ we shall denote by $L^p(\Omega, \mathscr{F}, \mathbb{P}; K)$ the space of all (equivalence class of) random variables $X \colon \Omega \to K$ such that

$$\int_\Omega |X(\omega)|^p \mathbb{P}(d\omega) < +\infty.$$

$L^p(H, \mathscr{B}(H), \mu; K)$, endowed with the norm

$$\|X\|_{L^p(\Omega,\mathscr{F},\mathbb{P};K)} = \left(\int_\Omega |X(\omega)|^p \mu(d\omega)\right)^{1/p},$$

is a Banach space.

Let $X \in L^2(\Omega, \mathscr{F}, \mathbb{P}; K)$. Let us compute the mean m_X, the covariance Q_X and the Fourier transform $\widehat{\mu_X}$ of the law $X_\#\mathbb{P}$ of X.

For any $h \in K$ we have, by (1.1)

$$\langle m_X, h\rangle = \int_K \langle y, h\rangle X_\#\mathbb{P}(dy) = \int_\Omega \langle X(\omega), h\rangle \mathbb{P}(d\omega). \tag{1.16}$$

In a similar way, for any $h, k \in K$, we have

$$\langle Q_X h, k \rangle = \int_K \langle y - m_X, h \rangle \langle y - m_X, k \rangle X_\# \mathbb{P}(dy)$$

(1.17)

$$= \int_\Omega \langle X(\omega) - m_X, h \rangle \langle X(\omega) - m_X, k \rangle \mathbb{P}d(\omega).$$

Finally,

$$\widehat{\mu_X}(k) = \int_K e^{i\langle y, k \rangle} X_\# \mathbb{P}(dy) = \int_\Omega e^{i\langle X(\omega), k \rangle} \mathbb{P}d(\omega).$$

Assume now that $X \in L^2(\Omega, \mathscr{F}, \mathbb{P}; \mathbb{R})$ is a real Gaussian random variable with law N_λ. Then for any $m \in \mathbb{N}$ we have

$$\int_\Omega |X(\omega)|^{2m} \mathbb{P}(d\omega) = (2\pi\lambda)^{-1/2} \int_{-\infty}^{+\infty} \xi^{2m} e^{-\frac{\xi^2}{2\lambda}} d\xi = \frac{(2m)!}{2^m \, m!} \lambda^m.$$

Therefore $X \in L^{2m}(\Omega, \mathscr{F}, \mathbb{P}; \mathbb{R})$ for all $m \in \mathbb{N}$.

Proposition 1.16 *Let (X_n) be a sequence of Gaussian random variables in a probability space $(\Omega, \mathscr{F}, \mathbb{P})$ with values in the separable Hilbert space K. Assume that for any $n \in \mathbb{N}$, X_n has mean a_n and covariance operators Q_n and that $X_n \to X$ in $L^2(\Omega, \mathscr{F}, \mathbb{P}; K)$. Then X is a Gaussian random variable with law $N_{a,Q}$ where*

$$\langle a, h \rangle = \lim_{n \to \infty} \langle a_n, h \rangle, \quad h \in K,$$

and

$$\langle Qh, k \rangle = \lim_{n \to \infty} \langle Q_n h, k \rangle, \quad h, k \in K.$$

Proof. Denote by a_n (resp. a) the mean and by Q_n (resp. Q) the covariance of $(X_n)_\# \mathbb{P}$ (resp. $X_\# \mathbb{P}$). We first notice that by the dominated convergence theorem we have for each $h, k \in K$,

$$\lim_{n \to \infty} \langle a_n, h \rangle = \lim_{n \to \infty} \int_\Omega \langle X_n(\omega), h \rangle \mathbb{P}(d\omega) = \int_\Omega \langle X(\omega), h \rangle \mathbb{P}(d\omega) = \langle a, h \rangle$$

and

$$\lim_{n \to \infty} \langle Q_n h, k \rangle = \lim_{n \to \infty} \int_\Omega \langle X_n(\omega) - a_n, h \rangle \langle X_n(\omega) - a_n, k \rangle \mathbb{P}(d\omega)$$

$$= \int_\Omega \langle X(\omega) - a, h \rangle \langle X(\omega) - a, k \rangle \mathbb{P}(d\omega) = \langle Qh, k \rangle,$$

for all $h, k \in K$.

Now, to show that X is Gaussian it is enough to prove (thanks to the uniqueness of Fourier transform) that

$$\int_H e^{i\langle y,k\rangle} X_\# \mathbb{P}(dx) = e^{i\langle a,k\rangle} e^{-\frac{1}{2}\langle Qk,k\rangle}, \quad k \in K.$$

We have in fact

$$\int_K e^{i\langle y,k\rangle} X_\# \mathbb{P}(dy) = \int_\Omega e^{i\langle X(\omega),k\rangle} \mathbb{P}(d\omega) = \lim_{n\to\infty} \int_\Omega e^{i\langle X_n(\omega),k\rangle} \mathbb{P}(d\omega)$$

$$= \lim_{n\to\infty} e^{i\langle a_n,k\rangle} e^{-\frac{1}{2}\langle Q_n k,k\rangle} = e^{i\langle a,k\rangle} e^{-\frac{1}{2}\langle Qk,k\rangle}.$$

Therefore X is Gaussian and $X_\# \mathbb{P} = N_{a,Q}$ as claimed. $\square$

1.6.1 Changes of variables involving Gaussian measures

Proposition 1.17 *Let $\mu = N_{a,Q}$ be a Gaussian measure on $(H, \mathscr{B}(H))$. Let $X(x) = x + b$, $x \in H$, where b is a fixed element of H. Then X is a Gaussian random variable and $X_\# \mu = N_{a+b,Q}$.*

Proof. We have in fact by (1.1)

$$\int_H e^{i\langle k,y\rangle} X_\# \mu(dy) = \int_H e^{i\langle x+b,k\rangle} \mu(dx) = e^{i\langle a+b,k\rangle} e^{-\frac{1}{2}\langle Qk,k\rangle}, \quad k \in H.$$

$\square$

Proposition 1.18 *Let $\mu = N_{a,Q}$ be a Gaussian measure on $(H, \mathscr{B}(H))$ and let $T \in L(H,K)$ where K is another Hilbert space. Then T is a Gaussian random variable and $T_\# \mu = N_{Ta,TQT^*}$, where T^* is the transpose of T.*

Proof. We have in fact by (1.1)

$$\int_K e^{i\langle k,y\rangle} T_\# \mu(dy) = \int_H e^{i\langle k,Tx\rangle} \mu(dx)$$

$$= \int_H e^{i\langle T^*k,x\rangle} \mu(dx) = e^{i\langle T^*k,a\rangle} e^{-\frac{1}{2}\langle TQT^*k,k\rangle}, \quad k \in K.$$

$\square$

Corollary 1.19 *Let $\mu = N_Q$ be a Gaussian measure on $(H, \mathscr{B}(H))$ and let $z_1, \ldots, z_k \in H$. Let $T \colon H \to \mathbb{R}^n$ defined as*

$$T(x) = (\langle x, z_1\rangle, \ldots, \langle x, z_n\rangle), \quad x \in H.$$

Then T is a Gaussian random variable $N_{Q'}$ with values in $\mathbb{R}^n$ where

$$Q'_{i,j} = \langle Qz_i, z_j\rangle, \quad i, j = 1, \ldots, n.$$

Proof. By a direct computation we see that the transpose $T^*: \mathbb{R}^n \to H$ of T is given by

$$T^*(\xi) = \sum_{i=1}^{n} \xi_i z_i, \quad \xi = (\xi_1, \ldots, \xi_n) \in \mathbb{R}^n.$$

Consequently we have

$$TQT^*(\xi) = T \sum_{i=1}^{k} \xi_i(Q z_i) = \sum_{i=1}^{k} \xi_i \left(\langle Q z_i, z_1 \rangle, \ldots, \langle Q z_i, z_k \rangle \right),$$

and the conclusion follows. $\square$

1.6.2 Independence

Let $(\Omega, \mathscr{F}, \mathbb{P})$ be a probability space and let $X_1, \ldots, X_n$ be real random variables in H. Consider the random variable X with values in $\mathbb{R}^n$,

$$X(\omega) = (X_1(\omega), \ldots, X_n(\omega)), \quad \omega \in H.$$

We say that $X_1, \ldots, X_n$ are *independent* if

$$X_\# \mathbb{P} = \bigtimes_{j=1}^{n} (X_j)_\# \mathbb{P}.$$

Proposition 1.20 *Let $X_1, \ldots, X_k$ be real independent random variables in $(\Omega, \mathscr{F}, \mathbb{P})$, $k \in \mathbb{N}$. Let moreover $\varphi_1, \ldots, \varphi_k$ be Borel real positive functions. Then we have*

$$\int_\Omega \varphi_1(X_1(\omega)) \cdots \varphi_k(X_k(\omega)) \mathbb{P}(d\omega)$$

$$= \int_\Omega \varphi_1(X_1(\omega)) \mathbb{P}(d\omega) \cdots \int_\Omega \varphi_k(X_k(\omega)) \mathbb{P}(d\omega). \tag{1.18}$$

Conversely, if (1.18) holds for any choice of $\varphi_1, \ldots, \varphi_k$ positive Borel, then $X_1, \ldots, X_k$ are independent.

Proof. Set $X = (X_1, \ldots, X_k)$ and let $\psi: \mathbb{R}^k \to \mathbb{R}$ be defined as

$$\psi(\xi_1, \ldots, \xi_k) = \varphi_1(\xi_1) \cdots \varphi_k(\xi_k), \quad (\xi_1, \ldots, \xi_k) \in \mathbb{R}^k.$$

Then by the change of variable formula (1.1) we have, taking into account the independence of $X_1, \ldots, X_k$,

$$\int_\Omega \varphi_1(X_1(\omega)) \cdots \varphi_k(X_k(\omega)) \mathbb{P}(d\omega) = \int_\Omega \psi(X(\omega)) \mathbb{P}(d\omega)$$

$$= \int_{\mathbb{R}^k} \psi(\xi) X_\# \mathbb{P}(d\xi) = \int_\mathbb{R} \varphi_1(\xi_1)(X_1)_\# \mathbb{P}(d\xi_1) \cdots \int_\mathbb{R} \varphi_k(\xi_k)(X_k)_\#(d\xi_k)$$

$$= \int_\Omega \varphi_1(X_1(\omega)) \mathbb{P}(d\omega) \cdots \int_\Omega \varphi_k(X_k(\omega)) \mathbb{P}(d\omega).$$

Assume conversely that (1.18) holds for any choice of $\varphi_1, \ldots, \varphi_k$ positive Borel. It is enough to show that

$$X_\# \mathbb{P}(I_1 \times \cdots \times I_k) = (X_1)_\# \mathbb{P} \cdots (X_k)_\# \mathbb{P} \quad \text{for all } I_1, \ldots, I_k \in \mathscr{B}(\mathbb{R}).$$

But this follows immediately setting in (1.18)

$$\varphi_i(\xi_i) = \mathbf{1}_{I_i}, \quad i = 1, \ldots, k.$$

$\square$

Exercise 1.21 Let X_1 and X_2 be independent real Gaussian random variables with laws N_{a_1, λ_1} and N_{a_2, λ_2} respectively. Show that $X_1 + X_2$ is a real Gaussian random variable with law $N_{a_1 + a_2, \lambda_1 + \lambda_2}$

Example 1.22 Let $\Omega = H$, $\mu = N_Q$, (e_k) an orthonormal basis in H and (λ_k) a sequence of positive numbers such that

$$Qe_k = \lambda_k e_k, \quad k \in \mathbb{N}.$$

Let $X_1, \ldots, X_n$ be defined by

$$X_1(x) = \langle x, e_1 \rangle, \ldots, X_n(x) = \langle x, e_n \rangle, \quad x \in H.$$

Then, by Corollary 1.19 we see that

$$(X_j)_\# \mu = N_{\langle Qe_j, e_j \rangle} = N_{\lambda_j},$$

and

$$X_\# \mu = N_{(Q_{i,j})},$$

where

$$Q_{i,j} = \langle Qe_i, e_j \rangle = \lambda_j \delta_{i,j}, \quad i, j = 1, \ldots, n.$$

Therefore we have

$$X_\# \mu = \underset{j=1}{\overset{n}{\times}} N_{\lambda_j} = \underset{j=1}{\overset{n}{\times}} (X_j)_\# \mu,$$

so that $X_1, \ldots, X_n$ are independent.

Example 1.23 Let $H = \mathbb{R}^n$, $Q = (Q_{i,j})_{i,j=1,\ldots,n} \in L_1^+(\mathbb{R}^n)$ such that $\det Q > 0$ and $\mu = N_Q$, so that

$$\mu(dx) = \frac{1}{\sqrt{(2\pi)^n \det Q}} \, e^{-\frac{1}{2}\langle Q^{-1}x,x\rangle} dx, \quad x \in \mathbb{R}^n.$$

Let $X_1,\ldots,X_n$ be random variables defined as

$$X_1(x) = x_1,\ldots,X_n(x) = x_n, \quad x = (x_1,\ldots,x_n) \in \mathbb{R}^n$$

and set

$$X = (X_1,\ldots,X_n).$$

By Corollary 1.19 we have

$$(X_j)_{\#}\mu = N_{Q_{j,j}}, \quad j = 1,\ldots,n,$$

and

$$X_{\#}\mu = N_{(Q_{i,j})}.$$

Therefore $X_1,\ldots,X_k$ are independent if and only if the matrix Q is diagonal.

Proposition 1.24 *Let $\mu = N_Q$ be a Gaussian measure on $(H,\mathscr{B}(H))$ and let $f_1,\ldots,f_n \in H$. Set*

$$X_{f_1}(x) = \langle x, f_1\rangle,\ldots,X_{f_n}(x) = \langle x, f_n\rangle, \quad x \in H,$$

and $X = (X_{f_1},\ldots,X_{f_n})$. Then $X_{f_1},\ldots,X_{f_n}$ are independent if and only if

$$\langle Qf_i, f_j\rangle = 0 \quad \text{if } f_i \neq f_j. \tag{1.19}$$

Proof. By Corollary 1.19 we see that

$$\mu = N_{(Q_{i,j})}, \quad Q_{i,j} = \langle Qf_i, f_j\rangle, \quad i,j = 1,\ldots,n,$$

and

$$(X_{f_j})_{\#}\mu = N_{\langle Qf_j, f_j\rangle}, \quad j = 1,\ldots,n.$$

Therefore, $X_{\#}\mu = \bigtimes_{i=1}^{n}(X_i)_{\#}\mu$ if and only if (1.19) holds. $\square$

Notice that if $f_1,\ldots,f_n$ are mutually orthogonal the random variables $X_{f_1},\ldots,X_{f_n}$ are not necessarily independent.

We end this section by proving that any nondegenerate Gaussian measure μ is *full*. This means that $\mu(D) > 0$ for any open non-empty subset D of H or, equivalently, that the support of μ is the whole H [1]

---

[1] We recall that the *support* of a probability measure μ in $(H,\mathscr{B}(H))$ is the intersection of all closed subsets of Ω of probability 1.

Proposition 1.25 Let $\mu = N_{a,Q}$ be a nondegenerate Gaussian measure in H. Then μ is full.

Proof. It is enough to show that any ball $B(x, r)$ of centre x and radius r has a positive probability. Let us fix $r > 0$ and take $x = 0$ for simplicity. Write for any $n \in \mathbb{N}$,

$$A_n = \left\{ x \in H : \sum_{k=1}^{n} x_k^2 \le \frac{r^2}{2} \right\}, \quad B_n = \left\{ x \in H : \sum_{k=n+1}^{\infty} x_k^2 < \frac{r^2}{2} \right\}.$$

Then we have

$$\mu(B(0, r)) \ge \mu(A_n \cap B_n) = \mu(A_n)\mu(B_n),$$

because A_n and B_n are independent (recall Example 1.22). Since clearly $\mu(A_n) > 0$ it is enough to show that $\mu(B_n) > 0$, provided n is sufficiently large. We have in fact by the Chebyshev inequality,

$$\mu(B_n) = 1 - \mu(B_n^c) \ge 1 - \frac{2}{r^2} \sum_{k=n+1}^{\infty} \int_H x_k^2 \mu(dx)$$

$$= 1 - \frac{2}{r^2} \sum_{k=n+1}^{\infty} (\lambda_k + a_k^2) > 0,$$

if n is large. $\square$

1.7 The Cameron–Martin space and the white noise mapping

In this section we consider a separable *infinite dimensional* Hilbert space H and a nondegenerate Gaussian measure $\mu = N_Q$ (that is such that Ker $Q = \{0\}$). We denote by (e_k) a complete orthonormal system on H such that $Q e_k = \lambda_k e_k$, $k \in \mathbb{N}$, where (λ_k) are the eigenvalues of Q. We set $x_k = \langle x, e_k \rangle$, $k \in \mathbb{N}$.

We notice that Q^{-1} is not continuous, since

$$Q^{-1} e_k = \frac{1}{\lambda_k} e_k, \quad k \in \mathbb{N}$$

and $\lambda_k \to 0$ as $k \to \infty$.

Let us now assume that $z_1, \ldots, z_n \in Q^{1/2}(H)$ and consider the linear functionals,

$$W_{z_i}(x) = \langle x, Q^{-1/2} z_i \rangle, \quad i = 1, \ldots, n.$$

Proposition 1.26 Let $z_1, \ldots, z_n \in H$. Then the law of the random variable (with values in $\mathbb{R}^n$) $(W_{z_1}, \ldots, W_{z_n})$ is given by

$$(W_{z_1}, \ldots, W_{z_n})_{\#}\mu = N_{(\langle z_i, z_j \rangle)_{i,j=1,\ldots,n}}. \tag{1.20}$$

Moreover, the random variables $W_{z_1}, \ldots, W_{z_n}$ are independent if and only if $z_1, \ldots, z_n$ is an orthogonal system, that is if and only if

$$\langle z_i, z_j \rangle = 0 \quad \text{for all } i \neq j, \ i, j = 1, \ldots, n. \tag{1.21}$$

Proof. Since

$$W_{z_i}(x) = F_{Q^{-1/2}z_i}(x), \quad x \in H, \ i = 1, \ldots, n,$$

the first statement follows from Corollary 1.19 and the second one from Proposition 1.24. $\square$

We want now to define a random variable W_z on $(H, \mathscr{B}(H), \mu)$ for all $z \in H$ such that (1.20) holds for all $z_1, \ldots, z_n \in H$. This random variable will be very useful in what follows, in particular in defining the Brownian motion. A first idea would be to define W_z by

$$W_z(x) = \langle Q^{-1/2}x, z \rangle, \quad x \in Q^{1/2}(H).$$

However this definition does not produce a random variable in H since $Q^{1/2}(H)$ is a μ-null set, as the following proposition shows.

Proposition 1.27 We have $\mu(Q^{1/2}(H)) = 0$.

Proof. For any $n, k \in \mathbb{N}$ set

$$U_n = \left\{ y \in H : \sum_{h=1}^{\infty} \lambda_h^{-1} y_h^2 < n^2 \right\},$$

and

$$U_{n,k} = \left\{ y \in H : \sum_{h=1}^{2k} \lambda_h^{-1} y_h^2 < n^2 \right\}.$$

Clearly $U_n \uparrow Q^{1/2}(H)$ as $n \to \infty$, and for any $n \in \mathbb{N}$, $U_{n,k} \downarrow U_n$ as $k \to \infty$. So it is enough to show that

$$\mu(U_n) = \lim_{k \to \infty} \mu(U_{n,k}) = 0. \tag{1.22}$$

We have in fact

$$\mu(U_{n,k}) = \int_{\left\{ y \in \mathbb{R}^k : \sum_{h=1}^{2k} \lambda_h^{-1} y_h^2 < n^2 \right\}} \bigtimes_{h=1}^{2k} N_{\lambda_k}(dy_k),$$

which, setting $z_h = \lambda_h^{-1/2} y_h$, is equivalent to

$$\mu(U_{n,k}) = \int_{\{z \in \mathbb{R}^{2k} : |z| < n\}} N_{I_{2k}}(dz),$$

where I_{2k} is the identity in $\mathbb{R}^{2k}$. Let us compute $\mu(U_{n,k})$. We have

$$\mu(U_{n,k}) = \frac{\mu(U_{n,k})}{\mu(H)} = \frac{\int_0^n e^{-\frac{r^2}{2}} r^{2k-1} dr}{\int_0^{+\infty} e^{-\frac{r^2}{2}} r^{2k-1} dr} = \frac{\int_0^{n^2/2} e^{-\rho} \rho^{k-1} d\rho}{\int_0^{+\infty} e^{-\rho} \rho^{k-1} d\rho}.$$

Therefore

$$\mu(U_{n,k}) = \frac{1}{(k-1)!} \int_0^{n^2/2} e^{-\rho} \rho^{k-1} d\rho \le \frac{1}{(k-1)!} \int_0^{n^2/2} \rho^{k-1} d\rho$$

$$= \frac{1}{k!} \left(\frac{n^2}{2} \right)^k,$$

and (1.22) follows. $\square$

$Q^{1/2}(H)$ is called the *Cameron–Martin space*.

Let now extend the definition on W_z for all $z \in H$ as follows. Consider the mapping W,

$$W : Q^{1/2}(H) \subset H \to L^2(H, \mu), \quad z \mapsto W_z, \quad W_z(x) = \langle x, Q^{-1/2}z \rangle, \quad x \in H.$$

If $z_1, z_2 \in Q^{1/2}(H)$ we have

$$\int_H W_{z_1}(x) W_{z_2}(x) \mu(dx) = \langle QQ^{-1/2}z_1, Q^{-1/2}z_2 \rangle = \langle z_1, z_2 \rangle \quad (1.23)$$

and so the mapping W is an isometry. Since $Q^{1/2}(H)$ is dense in H the mapping W can be uniquely extended to H. It is called the *white noise mapping*.

Proposition 1.28 *Let* $n \in \mathbb{N}$, $z_1, \ldots, z_n \in H$. *Then the law of* $(W_{z_1}, \ldots, W_{z_n})$ *in* $\mathbb{R}^n$ *is given by*

$$N_{(\langle z_i, z_j \rangle)_{i,j=1,\ldots,n}}. \quad (1.24)$$

The random variables $W_{z_1}, \ldots, W_{z_n}$ *are independent if and only if* $z_1, \ldots, z_n$ *are mutually orthogonal.*

Proof. Let

$$P_N x = \sum_{i=1}^N \langle x, e_i \rangle e_i, \quad x \in H, \ N \in \mathbb{N}.$$

Then, since W is an isomorphism, we have

$$\lim_{N\to\infty} (W_{P_N z_1}, \ldots, W_{P_N z_n}) = (W_{z_1}, \ldots, W_{z_n}) \quad \text{in } L^2(H, \mu; \mathbb{R}^n).$$

But by Corollary 1.19 it follows that $(W_{P_N z_1}, \ldots, W_{P_N z_n})$ is a random variable in $\mathbb{R}^n$ with law $N_{Q'_N}$ with

$$Q'_N = (\langle P_N z_i, P_N z_j \rangle)_{i,j=1,\ldots,n}.$$

So, by Proposition 1.16 we can conclude that $(W_{z_1}, \ldots, W_{z_n})$ is a random variable in $\mathbb{R}^n$ with law $N_{Q'}$ with

$$Q' = (\langle z_i, z_j \rangle)_{i,j=1,\ldots,n}.$$

The conclusion follows. $\square$

Exercise 1.29 Let $f \in H$. Prove that

$$\int_H e^{W_f(x)} \mu(dx) = e^{\frac{1}{2}|f|^2}. \tag{1.25}$$

Finally, we consider the function $f \to e^{W_f}$.

Proposition 1.30 *The mapping*

$$H \to L^2(H, \mu), \ f \mapsto e^{W_f}$$

is continuous.

Proof. Taking into account (1.25) it follows that

$$\int_H \left[e^{W_f} - e^{W_g} \right]^2 dN_Q = \int_H \left[e^{2W_f} - 2e^{W_{f+g}} + e^{2W_g} \right] dN_Q$$

$$= e^{2|f|^2} - 2e^{\frac{1}{2}|f+g|^2} + e^{2|g|^2}$$

$$= \left[e^{|f|^2} - e^{|g|^2} \right]^2 + 2e^{|f|^2+|g|^2} \left[1 - e^{-\frac{1}{2}|f-g|^2} \right],$$

which shows continuity in f of e^{W_f}. $\square$

The Cameron–Martin formula

2.1 Introduction and setting of the problem

We are given an infinite dimensional separable Hilbert space H and a nondegenerate Gaussian measure $\mu = N_Q$, where $Q \in L_1^+(H)$. We denote by (e_k) a complete orthonormal system on H such that $Qe_k = \lambda_k e_k$, $k \in H$, where (λ_k) are the eigenvalues of Q and set $x_k = \langle x, e_k \rangle$, $k \in \mathbb{N}$.

The main goal of this chapter is to study, following [11], equivalence and singularity of measures N_Q and $N_{a,Q}$, where $a \in H$.

We recall that if μ and ν are measures on $(\Omega, \mathscr{F})$ we say that μ is *absolutely continuous* with respect to ν (and we write $\mu \ll \nu$) if for all sets $A \in \mathscr{F}$ such that $\nu(A) = 0$ we have $\mu(A) = 0$.

If $\mu \ll \nu$ then by the Radon–Nikodym theorem, see e.g. [2], there exists a unique function $\rho \in L^1(\Omega, \mathscr{F}, \nu)$ such that

$$\mu(A) = \int_A \rho \, d\nu \quad \text{for all } A \in \mathscr{F}.$$

If $\mu \ll \nu$ and $\nu \ll \mu$ we say that μ and ν are *equivalent*.

Let first H be finite dimensional. Then, since Q is nondegenerate we have $\det Q > 0$. So, N_Q and $N_{a,Q}$ are equivalent and we have

$$\frac{dN_{a,Q}}{dN_Q}(x) = \frac{e^{-\frac{1}{2}\langle Q^{-1}(x-a), x-a \rangle}}{e^{-\frac{1}{2}\langle Q^{-1}x, x \rangle}} = e^{-\frac{1}{2}|Q^{-1/2}a|^2 + \langle Q^{-1/2}a, Q^{-1/2}x \rangle}, \quad x \in H.$$

$$(2.1)$$

We shall prove that

(i) If $a \in Q^{1/2}(H)$ then $N_{a,Q}$ and N_Q are equivalent.
(ii) If $a \notin Q^{1/2}(H)$ then $N_{a,Q}$ and N_Q are singular.

Moreover, in the first case we shall prove that formula (2.1) still holds, provided we interpret the term $\langle Q^{-1/2}a, Q^{-1/2}x \rangle$ as $W_{Q^{-1/2}a}(x)$.

We shall need a theorem of *Kakutani* about equivalence and singularity of products of infinitely many measures, proved in section 2.2. Section 2.3 is devoted to the proof of the Cameron–Martin theorem. Finally, in section 2.4 we shall study equivalence of two Gaussian measures N_Q and N_R with $Q, R \in L_1^+(H)$ (*Feldman–Hajek theorem*) in the particular case when Q and R commute.

2.2 Equivalence and singularity of product measures

Let us start with the notion of *Hellinger integral*. Let μ and ν be two probability measures on $(\Omega, \mathscr{F})$. It is obvious that μ and ν are absolutely continuous with respect to the probability measure $\zeta :=$ $\frac{1}{2}(\mu + \nu)$ on $(\Omega, \mathscr{F})$. Then the *Hellinger integral* of μ and ν is defined by

$$H(\mu, \nu) = \int_\Omega \sqrt{\frac{d\mu}{d\zeta} \frac{d\nu}{d\zeta}} \, d\zeta.$$

Notice that $0 \le H(\mu, \nu) \le 1$. In fact, by the Hölder inequality, we have

$$H(\mu, \nu) \le \left(\int_\Omega \frac{d\mu}{d\zeta} \, d\zeta \right)^{1/2} \left(\int_\Omega \frac{d\nu}{d\zeta} \, d\zeta \right)^{1/2} = 1.$$

Remark 2.1 Let λ be a probability measure on $(\Omega, \mathscr{F})$ such that $\mu \ll \lambda$ and $\nu \ll \lambda$. Then we have obviously $\zeta \ll \lambda$ and consequently,

$$\frac{d\mu}{d\zeta} = \frac{d\mu}{d\lambda} \frac{d\lambda}{d\zeta} \,, \qquad \frac{d\nu}{d\zeta} = \frac{d\nu}{d\lambda} \frac{d\lambda}{d\zeta} \,.$$

Therefore in this case $H(\mu, \nu)$ can also be written as

$$H(\mu, \nu) = \int_\Omega \sqrt{\frac{d\mu}{d\lambda} \frac{d\nu}{d\lambda}} \, d\lambda.$$

Remark 2.2 Assume that μ and ν are equivalent. Then we have

$$\frac{d\mu}{d\zeta} \frac{d\nu}{d\zeta} = \frac{d\mu}{d\zeta} \frac{d\nu}{d\mu} \frac{d\mu}{d\zeta} = \left[\frac{d\mu}{d\zeta} \right]^2 \frac{d\nu}{d\mu}$$

and therefore

$$H(\mu, \nu) = \int_\Omega \sqrt{\frac{d\nu}{d\mu} \frac{d\mu}{d\zeta}} \, d\zeta = \int_\Omega \sqrt{\frac{d\nu}{d\mu}} \, d\mu.$$

Example 2.3 Let $\Omega = \mathbb{R}$, $\mu = N_\lambda$, $\nu = N_{a,\lambda}$, $a \in \mathbb{R}$, $\lambda > 0$, then we have

$$\frac{d\nu}{d\mu}(x) = e^{-\frac{a^2}{2\lambda} + \frac{ax}{\lambda}}, \quad x \in \mathbb{R},$$

and so

$$H(\mu, \nu) = e^{-\frac{a^2}{4\lambda}} \int_{\mathbb{R}} e^{\frac{ax}{2\lambda}} N_\lambda(dx) = e^{-\frac{a^2}{8\lambda}}.$$

Let us prove an important property of the Hellinger integral.

Proposition 2.4 *Assume that $H(\mu, \nu) = 0$. Then μ and ν are singular.*

Proof. Let us denote by f and g the Radon–Nikodym derivatives

$$f = \frac{d\mu}{d\zeta}, \quad g = \frac{d\nu}{d\zeta},$$

where $\zeta = \frac{1}{2}(\mu + \nu)$. Since

$$H(\mu, \nu) = \int_\Omega \sqrt{fg}\, d\zeta = 0,$$

we have that $fg = 0$, ζ-a.e. Set

$$A = \{\omega \in \Omega : f(\omega) = 0\},$$

$$B = \{\omega \in \Omega : g(\omega) = 0\},$$

$$C = \{\omega \in \Omega : f(\omega)g(\omega) = 0\}.$$

Then we have $\zeta(C) = 1$ so that $\mu(C) = \nu(C) = 1$. Moreover,

$$\mu(A) = \int_A f\, d\zeta = 0, \quad \nu(B) = \int_B g\, d\zeta = 0.$$

Consequently, μ is concentrated on $B \backslash A$ and ν on $A \backslash B$. Therefore, μ and ν are singular. $\square$

If $H(\mu, \nu) > 0$ the measures μ and ν are not necessarily equivalent in general. However, this happens for the product of equivalent measures, see Theorem 2.7 below. We first need a lemma.

Lemma 2.5 *Let $\mu_1, \nu_1, \mu_2, \nu_2$ be probability measures on $(\Omega, \mathscr{F})$. Then we have*

$$H(\mu_1 \times \mu_2, \nu_1 \times \nu_2) = H(\mu_1, \nu_1)H(\mu_2, \nu_2).$$

Proof. Let ζ_1, ζ_2 be probability measures on $(\Omega, \mathscr{F})$ such that

$$\mu_1 \ll \zeta_1, \quad \nu_1 \ll \zeta_1, \quad \mu_2 \ll \zeta_2, \quad \nu_2 \ll \zeta_2.$$

Then by the Fubini theorem

$$\mu_1 \times \mu_2 \ll \zeta_1 \times \zeta_2, \quad \nu_1 \times \nu_2 \ll \zeta_1 \times \zeta_2.$$

Set

$$f_1(\omega_1) = \frac{d\mu_1}{d\zeta_1}(\omega_1), \quad g_1(\omega_1) = \frac{d\nu_1}{d\zeta_1}(\omega_1)$$

and

$$f_2(\omega_2) = \frac{d\mu_2}{d\zeta_2}(\omega_2), \quad g_2(\omega_2) = \frac{d\nu_2}{d\zeta_2}(\omega_2).$$

Consequently

$$\frac{d(\mu_1 \times \mu_2)}{d(\zeta_1 \times \zeta_2)}(\omega_1, \omega_2) = f_1(\omega_1)f_2(\omega_2), \quad \frac{d(\nu_1 \times \nu_2)}{d(\zeta_1 \times \zeta_2)}(\omega_1, \omega_2) = g_1(\omega_1)g_2(\omega_2)$$

so that,

$$H(\mu_1 \times \mu_2, \nu_1 \times \nu_2) = \int_{\Omega \times \Omega} \sqrt{f_1(\omega_1)g_1(\omega_1)f_2(\omega_2)g_2(\omega_2)} \; \zeta_1(d\omega_1)\zeta_2(d\omega_2)$$

$$= H(\mu_1, \nu_1)H(\mu_2, \nu_2).$$

$\square$

Exercise 2.6 Let (μ_k) and (ν_k) be sequences of probability measures on $(\mathbb{R}, \mathscr{B}(\mathbb{R}))$. Consider the product measures on $(\mathbb{R}^\infty, \mathscr{B}(\mathbb{R}^\infty))$,

$$\mu = \underset{k=1}{\overset{\infty}{\times}} \mu_k, \quad \nu = \underset{k=1}{\overset{\infty}{\times}} \nu_k.$$

Prove that

$$H(\mu, \nu) = \prod_{k=1}^{\infty} H(\mu_k, \nu_k).$$

We are now ready to prove the following result.

Theorem 2.7 (Kakutani) Let (μ_k) and (ν_k) be sequences of probability measures on $(\mathbb{R}, \mathscr{F}(\mathbb{R}))$ such that μ_k and ν_k are equivalent for all $k \in \mathbb{N}$, and let $\mu = \underset{k=1}{\overset{\infty}{\times}} \mu_k$, $\nu = \underset{k=1}{\overset{\infty}{\times}} \nu_k$. If $H(\mu, \nu) > 0$, then μ and ν are equivalent and we have

$$\frac{d\nu}{d\mu}(x) = \lim_{n \to \infty} \prod_{i=1}^{n} \frac{d\nu_k}{d\mu_k}(x_k) \quad in \; L^1(\mathbb{R}^\infty, \mu). \tag{2.2}$$

If $H(\mu, \nu) = 0$, then μ and ν are singular.

Proof. Write

$$\rho_i(x_i) = \frac{d\nu_i}{d\mu_i}(x_i), \quad f_n(x_1, \ldots, x_n) = \prod_{i=1}^{n} \rho_i(x_i), \quad i, n \in \mathbb{N},$$

and for any $n \in \mathbb{N}$ define the measures

$$\mu^{(n)} = \underset{k=1}{\overset{n}{\times}} \mu_k, \quad \nu^{(n)} = \underset{k=1}{\overset{n}{\times}} \nu_k, \quad n \in \mathbb{N}.$$

Then we have (see Remark 2.2),

$$H(\mu^{(n)}, \nu^{(n)}) = \int_{\mathbb{R}^\infty} \prod_{k=1}^{n} \sqrt{\rho_k(x_k)} \, \mu^{(n)}(dx).$$

Obviously $\mu^{(n)}$ is equivalent to $\nu^{(n)}$ and

$$\frac{d\mu^{(n)}}{d\nu^{(n)}}(x) = f_n(x), \quad n \in \mathbb{N}, \ x \in \mathbb{R}^\infty.$$

We claim that the sequence (f_n) is convergent in $L^1(H, \mu)$.

To prove the claim it is enough to show that the sequence $(\sqrt{f_n})$ is convergent in $L^2(H, \mu)$. If $n, p \in \mathbb{N}$ we have in fact, taking into account independence of the random variables $(x_n)_{n \in \mathbb{N}}$,

$$\int_{\mathbb{R}^\infty} |\sqrt{f_{n+p}} - \sqrt{f_n}|^2 d\mu = \int_{\mathbb{R}^\infty} \prod_{k=1}^{n} \rho_k(x_k) \left| \prod_{k=n+1}^{n+p} \sqrt{\rho_k(x_k)} - 1 \right|^2 \mu(dx)$$

$$= \int_{\mathbb{R}^\infty} \prod_{k=1}^{n} \rho_k(x_k) \mu(dx) \int_{\mathbb{R}^\infty} \left| \prod_{k=n+1}^{n+p} \sqrt{\rho_k(x_k)} - 1 \right|^2 \mu(dx)$$

$$= \int_{\mathbb{R}^\infty} \left| \prod_{k=n+1}^{n+p} \sqrt{\rho_k(x_k)} - 1 \right|^2 \mu(dx)$$

$$= \int_{\mathbb{R}^\infty} \left[\prod_{k=n+1}^{n+p} \rho_k(x_k) - 2 \prod_{k=n+1}^{n+p} \sqrt{\rho_k(x_k)} + 1 \right] \mu(dx)$$

$$= 2 \left(1 - \prod_{k=n+1}^{n+p} \int_{\mathbb{R}} \sqrt{\rho_k(x_k)} \mu(dx) \right) = 2 - \prod_{k=n+1}^{n+p} H(\mu_k, \nu_k). \quad (2.3)$$

On the other hand, we know by assumption that

$$H(\mu, \nu) = \prod_{k=1}^{\infty} H(\mu_k, \nu_k) > 0,$$

or, equivalently, that

$$-\log H(\mu, \nu) = -\sum_{k=1}^{\infty} \log [H(\mu_k, \nu_k)] < +\infty.$$

Consequently, for any $\varepsilon > 0$ there exists $n_\varepsilon \in \mathbb{N}$ such that if $n > n_\varepsilon$ and $p \in \mathbb{N}$, we have

$$-\sum_{k=n+1}^{n+p} \log [H(\mu_k, \nu_k)] < \varepsilon.$$

By (2.3) it follows that for all $n > n_\varepsilon$

$$\int_{\mathbb{R}^\infty} |\sqrt{f_{n+p}} - \sqrt{f_n}|^2 d\mu \le 2(1 - e^{-\varepsilon}),$$

and the claim is proved.

We can now conclude the proof. Let f be the limit of (f_n) in $L^1(H, \mu)$. It is enough to prove that f coincides with the density $\frac{d\nu}{d\mu}$. Let φ be a real Borel function on $(\mathbb{R}^\infty, \mathscr{B}(\mathbb{R}^\infty))$ depending only on $x_1, \ldots, x_k$. Then for any $n > k$ we have

$$\int_{\mathbb{R}^\infty} \varphi(x) \nu^{(n)}(dx) = \int_{\mathbb{R}^\infty} \varphi(x) f_n(x) \mu^{(n)}(dx).$$

If $n > k$ the identity above is equivalent to

$$\int_{\mathbb{R}^\infty} \varphi(x) \nu(dx) = \int_{\mathbb{R}^\infty} \varphi(x) f_n(x) \mu(dx).$$

As $n \to \infty$ we obtain finally

$$\int_{\mathbb{R}^\infty} \varphi(x) \nu(dx) = \int_{\mathbb{R}^\infty} \varphi(x) f(x) \mu(dx).$$

This proves that $\nu \ll \mu$ in view of the arbitrariness of k and φ. Similarly we have $\mu \ll \nu$. Finally, the last statement follows from Proposition 2.4. □

2.3 The Cameron–Martin formula

Here we consider two Gaussian measures $\mu = N_Q$ and $\nu = N_{a,Q}$ on $(H, \mathscr{B}(H))$, where $a \in H$ and $Q \in L_1^+(H)$.

Theorem 2.8 *(i) If $a \notin Q^{1/2}(H)$ then μ and ν are singular.*
(ii) If $a \in Q^{1/2}(H)$ then μ and ν are equivalent.
(iii) If μ and ν are equivalent the density $\frac{d\nu}{d\mu}$ is given by

$$\frac{d\nu}{d\mu}(x) = \exp\left\{-\frac{1}{2}|Q^{-1/2}a|^2 + W_{Q^{-1/2}a}(x)\right\}, \quad x \in H. \quad (2.4)$$

Proof. Let us prove (i). Assume that $a \notin Q^{1/2}(H)$. Then $H(\mu, \nu) = 0$ and, by the Kakutani theorem it follows that μ and ν are singular.

Let us prove (ii). Notice first that, recalling Example 2.3, we have

$$H(\mu, \nu) = \prod_{k=1}^{\infty} H(\mu_k, \nu_k) = \prod_{k=1}^{\infty} e^{-\frac{a_k^2}{8\lambda_k}},$$

which implies that

$$-\log H(\mu, \nu) = \frac{1}{8}\sum_{k=1}^{\infty} \frac{a_k^2}{\lambda_k} = \begin{cases} \frac{1}{8}|Q^{-1/2}a|^2 & \text{if } a \in Q^{1/2}(H), \\[2ex] +\infty & \text{otherwise.} \end{cases}$$

Now, if $a \in Q^{1/2}(H)$ we have

$$|Q^{-1/2}a|^2 = \sum_{k=1}^{\infty} \frac{a_k^2}{\lambda_k} < +\infty$$

and so $H(\mu, \nu) > 0$ and, again by the Kakutani theorem, μ and ν are equivalent and (ii) is proved.

It remains to prove (iii). Assume that $a \in Q^{1/2}(H)$. Then, by (2.2) we know that

$$\frac{d\nu}{d\mu}(x) = \lim_{n\to\infty} \prod_{k=1}^{n} e^{-\frac{1}{2}\frac{a_k^2}{\lambda_k} + \frac{a_k x_k}{\lambda_k}} \quad \text{in } L^1(H, \mu). \quad (2.5)$$

For any $n \in \mathbb{N}$ denote by P_n the othogonal projector on the span of $e_1, \ldots, e_n$ and set $Q_n = P_n Q$. Then we can write (2.5) as

$$\frac{d\nu}{d\mu}(x) = \lim_{n\to\infty} e^{-\frac{1}{2}|Q_n^{-1/2}a|^2 + \langle Q_n^{-1/2}a, Q_n^{-1/2}x\rangle}$$

$$= \lim_{n\to\infty} e^{-\frac{1}{2}|Q_n^{-1/2}a|^2 + W_{Q_n^{-1/2}a}(x)} \quad \text{in } L^1(H, \mu). \quad (2.6)$$

On the other hand, since $a \in Q^{1/2}(H)$ we have $\lim_{n\to\infty} Q_n^{-1/2}a = Q_a^{-1/2}$. So, recalling Proposition 1.30, we see that (2.4) is fulfilled. $\square$

2.4 The Feldman–Hajek theorem

We are given two nondegenerate Gaussian measures $\mu = N_Q$, and $\nu = N_R$ where $Q, R \in L_1^+(H)$. We are going to show that μ and ν are either singular or equivalent. For the sake of simplicity we only consider the special case when Q and R commute. For the general case see e.g. [11].

The main tools are, as before, the Hellinger integral and the Kakutani theorem. Since Q and R commute there exists a complete orthonormal system (e_k) in H, and sequences $(\lambda_k), (r_k)$ of positive numbers such that

$$Qe_k = \lambda_k e_k, \quad Re_k = r_k e_k, \quad k \in \mathbb{N}.$$

Theorem 2.9 *Let $Q, R \in L_1^+(H)$ be such that $[Q, R]: = QR - RQ = 0$. Let $\mu = N_Q$ and $\nu = N_R$. Then μ and ν are equivalent if and only if*

$$\sum_{k=1}^{\infty} \frac{(\lambda_k - r_k)^2}{(\lambda_k + r_k)^2} < \infty.$$

If μ and ν are not equivalent they are singular.

Proof. Write

$$N_Q = \underset{k=1}{\overset{\infty}{\times}} N_{\lambda_k}, \quad N_R = \underset{k=1}{\overset{\infty}{\times}} N_{r_k}.$$

In order to apply the Kakutani theorem, let us compute the Hellinger integral

$$H(\mu, \nu) = \prod_{k=1}^{\infty} H(\mu_k, \nu_k).$$

Since

$$\frac{d\nu_k}{d\mu_k}(x_k) = \sqrt{\frac{\lambda_k}{r_k}} \, e^{-\frac{x_k^2}{2}\left(\frac{\lambda_k - r_k}{r_k \lambda_k}\right)}, \quad x_k \in \mathbb{R}, \ k \in \mathbb{N},$$

we have

$$H(\nu_k, \mu_k) = \int_{\mathbb{R}} \sqrt{\frac{d\nu_k}{d\mu_k}(x)} \, \mu_k(dx) = \left[\frac{4r_k\lambda_k}{(r_k + \lambda_k)^2}\right]^{1/4} = \left[\frac{4\xi_k}{(1 + \xi_k)^2}\right]^{1/4},$$

where $\xi_k = \frac{\lambda_k}{r_k}$. Consequently,

$$H^4(\mu, \nu) = \prod_{k=1}^{\infty} \frac{4\xi_k}{(1 + \xi_k)^2} = \prod_{k=1}^{\infty}\left(1 - \frac{(1 - \xi_k)^2}{(1 + \xi_k)^2}\right). \tag{2.7}$$

So, $H(\mu, \nu) > 0$ if and only if

$$\sum_{k=1}^{\infty} \frac{(1 - \xi_k)^2}{(1 + \xi_k)^2} = \sum_{k=1}^{\infty} \frac{(\lambda_k - r_k)^2}{(\lambda_k + r_k)^2} < \infty.$$

Then the conclusion follows from the Kakutani theorem. $\square$

Remark 2.10 Let $R = \alpha Q$ with $\alpha > 0$. Then by the previous theorem it follows that if $\alpha \neq 1$, N_Q and N_R are singular.

3

Brownian motion

Let $(\Omega, \mathscr{F}, \mathbb{P})$ be a probability space. An arbitrary family $X(t), t \geq 0$, of real random variables $X(t)$ defined on Ω is called a *stochastic process*. The functions $t \to X(t)(\omega)$, $\omega \in \Omega$ are called the *trajectories* of $X(t)$.

A stochastic process $Y(t)$ is said to be a *version* of $X(t)$ if

$$\mathbb{P}(X(t) \neq Y(t)) = 0 \quad \text{for all } t \geq 0.$$

A stochastic process $X(t)$ is said to be *measurable* if the mapping

$$[0, +\infty) \times \Omega \to \mathbb{R}, \ (t, \omega) \mapsto X(t)(\omega)$$

is $\mathscr{B}([0, +\infty)) \times \mathscr{F}$-measurable. $X(t)$ is said to be *continuous* if its trajectories are continuous for almost all $\omega \in \Omega$.

A real *Brownian motion* $B = (B(t))_{t \geq 0}$ is a continuous real stochastic process such that

(i) $B(0) = 0$ and if $0 \leq s < t$, $B(t) - B(s)$ is a real Gaussian random variable with law N_{t-s}.
(ii) If $0 < t_1 < \cdots < t_n$, the random variables

$$B(t_1), \ B(t_2) - B(t_1), \ldots, B(t_n) - B(t_{n-1})$$

are independent.

Property (ii) is expressed by saying that the Brownian motion has *independent increments*.

3.1 Construction of a Brownian motion

We shall construct a Brownian motion in the probability space $(H, \mathscr{B}(H), \mu)$ where $H = L^2(0, +\infty)$, $\mu = N_Q$, and Q is any operator in

$L_1^+(H)$ such that Ker $Q = \{0\}$. Let (e_n) be a complete orthonormal system in H and (λ_n) a sequence of positive numbers such that

$$Qe_n = \lambda_n e_n, \quad n \in \mathbb{N}.$$

Moreover, set

$$P_n x = \sum_{k=1}^{n} \langle x, e_k \rangle e_k, \quad x \in H, \ n \in \mathbb{N}.$$

Theorem 3.1 *Let* $B(t) = W_{\mathbf{1}_{[0,t]}}$, $t \geq 0$, *where*

$$\mathbf{1}_{[0,t]}(s) = \begin{cases} 1 & \text{if } s \in [0,t], \\ 0 & \text{otherwise}, \end{cases}$$

and W *is the white noise mapping. Then a version of* B *is a real Brownian motion on* $(H, \mathscr{B}(H), \mu)$.

Proof. Clearly $B(0) = 0$. Moreover, since for $t > s$,

$$B(t) - B(s) = W_{\mathbf{1}_{[0,t]}} - W_{\mathbf{1}_{[0,s]}} = W_{\mathbf{1}_{(s,t]}},$$

we know by Proposition 1.28 that $B(t) - B(s)$ is a real Gaussian random variable with law N_{t-s}, and (i) is proved. Let us prove (ii). Since the system of elements of H,

$$(\mathbf{1}_{[0,t_1]}, \mathbf{1}_{(t_1,t_2]}, \dots, \mathbf{1}_{(t_{n-1},t_n]}),$$

is orthogonal, we have again by Proposition 1.28 that the random variables $B(t_1)$, $B(t_2) - B(t_1), \dots, B(t_n) - B(t_{n-1})$ are independent. Thus (ii) is proved. It remains to show that almost all trajectories of (a version) of B are continuous. It is easy to see that $B(t)$ is measurable, [1] and that its trajectories belong to $L^{2m}(0,T)$ for all $m \in \mathbb{N}$, $T > 0$ and almost all $x \in H$. In fact, since the law of $B(t)$ is N_t we have

$$\int_H |B(t)(x)|^{2m} \mu(dx) = \int_{\mathbb{R}} |\xi|^{2m} N_t(d\xi) = \frac{(2m)!}{2^m m!} t^m,$$

so that, using the Fubini theorem,

$$\int_0^T \left[\int_H |B(t)(x)|^{2m} \mu(dx) \right] dt = \int_H \left[\int_0^T |B(t)(x)|^{2m} dt \right] \mu(dx) dt$$

$$= \frac{(2m)!}{(m+1)2^m m!} T^{m+1}.$$

This shows that the mapping $t \mapsto B(t)(x) \in L^{2m}(0,T)$, μ-a.e.

---

[1] It is enough to approximate $B(t)$ by $W_{P_n \mathbf{1}_{[0,T]}}$.

To show continuity of the trajectories, we shall represent $B(t)$ as the integral of a suitable L^{2m} function; this will yield continuity of $B(t)$ by an elementary analytic lemma. We shall use the *factorization method*, see e.g. [9]. It is based on the following elementary identity

$$\int_s^t (t-\sigma)^{\alpha-1}(\sigma-s)^{-\alpha}d\sigma = \frac{\pi}{\sin \pi\alpha}, \quad 0 \le s \le \sigma \le t, \qquad (3.1)$$

where $\alpha \in (0,1)$. To check (3.1) it is enough to set $\sigma = r(t-s) + s$ so that (3.1) becomes

$$\int_0^1 (1-r)^{\alpha-1}r^{-\alpha}dr = \frac{\pi}{\sin \pi\alpha},$$

which can easily be proved.

From now on we take $\alpha < 1/2$. Identity (3.1) can be written as

$$1_{[0,t]}(s) = \frac{\sin \pi\alpha}{\pi} \int_0^t (t-\sigma)^{\alpha-1}1_{[0,\sigma]}(s)(\sigma-s)^{-\alpha}d\sigma, \quad t > 0, \ s > 0.$$

We also write equivalently

$$1_{[0,t]} = \frac{\sin \pi\alpha}{\pi} \int_0^t (t-\sigma)^{\alpha-1}g_\sigma d\sigma, \qquad (3.2)$$

where

$$g_\sigma(s) = 1_{[0,\sigma]}(s)(\sigma-s)^{-\alpha}.$$

Since $\alpha < 1/2$, we have $g_\sigma \in H$ and $|g_\sigma|^2 = \frac{\sigma^{1-2\alpha}}{1-2\alpha}$. Recalling that the mapping

$$H \to L^2(H,\mu), \ f \mapsto W_f,$$

is continuous, we obtain the following representation formula for B,

$$B(t) = \frac{\sin \pi\alpha}{\pi} \int_0^t (t-\sigma)^{\alpha-1}W_{g_\sigma}d\sigma. \qquad (3.3)$$

Now it is enough to prove that the mapping $\sigma \mapsto W_{g_\sigma}(x)$ belongs to $L^{2m}(0,T)$, μ-a.e. for any $T > 0$; in fact this implies that B is continuous by Lemma 3.2 below.

To show $2m$-summability of the mapping $\sigma \mapsto W_{g_\sigma}(x)$ we notice that, since W_{g_σ} is a real Gaussian random variable with law $N_{\frac{\sigma^{1-2\alpha}}{1-2\alpha}}$, we have

$$\int_H |W_{g_\sigma}(x)|^{2m}\mu(dx) = \frac{(2m)!}{2^m m!}(1-2\alpha)^{-m}\sigma^{m(1-2\alpha)}. \qquad (3.4)$$

Since $\alpha < 1/2$ we have, by the Fubini theorem,

$$\int_0^T \left[\int_H |W_{g_\sigma}(x)|^{2m} \mu(dx) \right] d\sigma = \int_H \left[\int_0^T |W_{g_\sigma}(x)|^{2m} d\sigma \right] \mu(dx) < +\infty.$$
(3.5)

Therefore the mapping $\sigma \mapsto W_{g_\sigma}(x)$ belongs to $L^{2m}(0,T)$, μ-a.e. and the conclusion follows. $\square$

Lemma 3.2 *Let* $m > 1, \alpha \in (1/(2m), 1), T > 0$, *and* $f \in L^{2m}(0,T;H)$. *Set*

$$F(t) = \int_0^t (t-\sigma)^{\alpha-1} f(\sigma) d\sigma, \quad t \in [0,T].$$

Then $F \in C([0,T];H)$.

Proof. By Hölder's inequality we have (notice that $2m\alpha - 1 > 0$),

$$|F(t)| \le \left(\int_0^t (t-\sigma)^{(\alpha-1)\frac{2m}{2m-1}} d\sigma \right)^{\frac{2m-1}{2m}} |f|_{L^{2m}(0,T;H)}.$$
(3.6)

Therefore $F \in L^\infty(0,T;H)$. It remains to show continuity of F. Continuity at 0 follows from (3.6). Let us prove that F is continuous on $[\frac{t_0}{2}, T]$ for any $t_0 \in (0,T]$. Let us set for $\varepsilon < \frac{t_0}{2}$,

$$F_\varepsilon(t) = \int_0^{t-\varepsilon} (t-\sigma)^{\alpha-1} f(\sigma) d\sigma, \; t \in [0,T].$$

F_ε is obviously continuous on $[\frac{t_0}{2}, T]$. Moreover, using once again Hölder's estimate, we find

$$|F(t) - F_\varepsilon(t)| \le M \left(\frac{2m-1}{2m\alpha - 1} \right)^{\frac{2m-1}{2m}} \varepsilon^{\alpha - \frac{1}{2m}} |f|_{L^{2m}(0,T;H)}.$$

Thus $\lim_{\varepsilon \to 0} F_\varepsilon(t) = F(t)$, uniformly on $[\frac{t_0}{2}, T]$, and F is continuous as required. $\square$

Exercise 3.3 Prove that $\int_H B(t)B(s)d\mu = \min\{t,s\}$ for all $t, s \ge 0$.

Exercise 3.4 Prove that for any $T > 0$, $B(\cdot)x$ is Hölder continuous in $[0,T]$ with any exponent $\beta < 1/2$ for μ-almost all $x \in H$.

Exercise 3.5 Let B be a Brownian motion in a probability space $(\Omega, \mathscr{F}, \mathbb{P})$. Prove that the following are Brownian motions.

(i) Invariance by translation

$$B_1(t) = B(t+h) - B(h), \quad t \ge 0, \quad \text{where } h > 0 \text{ is given.}$$

(ii) Self-similarity,

$$B_2(t) = \alpha B(\alpha^{-2}t), \quad \text{where } \alpha > 0 \text{ is given.}$$

(iii) Time reversal

$$B_3(t) = tB(1/t), \quad t > 0, \; B_3(0) = 0.$$

(iv) Symmetry

$$B_4(t) = -B(t), \quad t \geq 0.$$

Let us now introduce the standard Brownian motion and the Wiener measure. Set

$$X = \{x \in H : B(\cdot, x) \text{ is continuous}\}.$$

Then the restriction of $B(t)$ to the probability space $(X, \mathscr{B}(H) \cap X, \mu)$ (which we still denote by $B(t)$) is a Brownian motion having all trajectories continuous.

Let now $C_0([0,T]) = \{\omega \in C([0,T]) : \omega(0) = 0\}$ and consider the mapping $B(\cdot)$

$$X \to C_0([0,T]), \; x \mapsto B(\cdot, x).$$

$B(\cdot)$ is clearly measurable from $(X, \mathscr{B}(H) \cap X, \mu)$ into $(C_0([0,T]), \mathscr{B}(C_0([0,T])))$. The law of $B(t)$ $\mathbb{P} := B(\cdot)_{\#}\mu$ is called the *Wiener measure* on $(C_0([0,T]), \mathscr{B}(C_0([0,T])))$. $\mathbb{P}$ is obviously defined by the change of variables formula

$$\int_X F(B(\cdot, x))\mu(dx) = \int_{C_0([0,T])} F(\omega)\mathbb{P}(d\omega). \tag{3.7}$$

Consider the following stochastic process W in $(C_0([0,T]), \mathscr{B}(C_0([0,T])), \mathbb{P})$

$$W(t)(\omega) = \omega(t), \quad \omega \in C_0([0,T]), \; t \geq 0.$$

Using formula (3.7) one can check easily that W is a Brownian motion in $(C_0([0,T]), (C_0([0,T])), \mathbb{P})$, called the *standard Brownian motion*.

3.2 Total variation of a Brownian motion

We are here concerned with a Brownian motion B in a probability space $(\Omega, \mathscr{F}, \mathbb{P})$. We are going to show that the *total variation* of $B(\cdot)(\omega)$ in $[0,T]$ is infinite for all $T > 0$ and $\mathbb{P}$-almost all $\omega \in \Omega$.

Let us denote by Σ the set of all decompositions

$$\sigma = \{0 = t_0 < t_1 < \cdots < t_n = T\}$$

of the interval $[0, T]$. For any $\sigma \in \Sigma$ we set

$$|\sigma| = \max_{k=1,\dots,n} (t_k - t_{k-1}).$$

The set Σ is endowed with the usual partial ordering

$$\sigma_1 < \sigma_2 \iff |\sigma_1| \le |\sigma_2|.$$

Let $f \in C([0, T])$. Let us recall that the *total variation* $V_T(f)$ of f in $[0, T]$ is defined as

$$V_T(f) = \sup_{\sigma \in \Sigma} \sum_{k=1}^{n} |f(t_k) - f(t_{k-1})|, \quad \sigma = \{0 = t_0 < t_1 < \cdots < t_n = T\}.$$

If $V_T(f) < +\infty$ we say that f has *bounded total variation* in $[0, T]$.

Let us also introduce the *quadratic variation* of f setting,

$$J(f) := \lim_{|\sigma| \to 0} \sum_{k=1}^{n} |f(t_k) - f(t_{k-1})|^2, \quad \sigma = \{0 = t_0 < t_1 < \cdots < t_n = T\},$$

whenever the limit above exists.

Exercise 3.6 Let $f \in C([0, T])$ and assume that $J(f) > 0$. Prove that $V_T(f) = +\infty$.

We shall prove first that $J(B(\cdot)) = T$, $\mathbb{P}$-a.e. Then, taking into account Exercise 3.6, it will follow easily that $V(B(\cdot)) = +\infty$, $\mathbb{P}$-a.e.

Lemma 3.7 *Set*

$$J_\sigma = \sum_{k=1}^{n} |B(t_k) - B(t_{k-1})|^2,$$

for $\sigma = \{0 = t_0 < t_1 < \cdots < t_n = T\} \in \Sigma$. *Then we have*

$$\lim_{|\sigma| \to 0} J_\sigma = T \quad in \ L^2(\Omega, \mathscr{F}, \mathbb{P}). \tag{3.8}$$

Proof. Let $\sigma = \{0 = t_0 < t_1 < \cdots < t_n = T\} \in \Sigma$. Then we have

$$\int_\Omega |J_\sigma - T|^2 d\mathbb{P} = \int_\Omega J_\sigma^2 d\mathbb{P} - 2T \int_\Omega J_\sigma d\mathbb{P} + T^2. \tag{3.9}$$

But

$$\int_\Omega J_\sigma d\mathbb{P} = \sum_{k=1}^{n} \int_\Omega |B(t_k) - B(t_{k-1})|^2 d\mathbb{P} = \sum_{k=1}^{n} (t_k - t_{k-1}) = T, \tag{3.10}$$

since $B(t_k) - B(t_{k-1})$ is a Gaussian random variable with law $N_{t_k - t_{k-1}}$. Moreover

$$\int_\Omega |J_\sigma|^2 d\mathbb{P} = \int_\Omega \left| \sum_{k=1}^n |B(t_k) - B(t_{k-1})|^2 \right|^2 d\mathbb{P}$$

$$= \int_\Omega \sum_{k=1}^n |B(t_k) - B(t_{k-1})|^4 d\mathbb{P}$$

$$+ 2 \sum_{h<k=1}^n \int_\Omega |B(t_h) - B(t_{h-1})|^2 |B(t_k) - B(t_{k-1})|^2 d\mathbb{P}.$$

Now, using again the fact that $B(t_k) - B(t_{k-1})$ is a Gaussian random variable with law $N_{t_k - t_{k-1}}$, we have

$$\int_\Omega \sum_{k=1}^n |B(t_k) - B(t_{k-1})|^4 d\mathbb{P} = 3 \sum_{k=1}^n (t_k - t_{k-1})^2, \qquad (3.11)$$

and, since $B(t_h) - B(t_{h-1})$ and $B(t_k) - B(t_{k-1})$ are independent,

$$\sum_{h<k=1}^n \int_\Omega |B(t_h) - B(t_{h-1})|^2 |B(t_k) - B(t_{k-1})|^2 d\mu$$

$$= \sum_{h<k=1}^n (t_h - t_{h-1})(t_k - t_{k-1}). \qquad (3.12)$$

Therefore

$$\int_\Omega |J_\sigma|^2 d\mathbb{P} = 3 \sum_{k=1}^n (t_k - t_{k-1})^2 + 2 \sum_{h<k=1}^n (t_h - t_{h-1})(t_k - t_{k-1})$$

$$= 2 \sum_{k=1}^n (t_k - t_{k-1})^2 + \left(\sum_{k=1}^n (t_k - t_{k-1}) \right)^2$$

$$= 2 \sum_{k=1}^n (t_k - t_{k-1})^2 + T^2. \qquad (3.13)$$

Now, substituting (3.10) and (3.13) in (3.9), we obtain

$$\int_\Omega |J_\sigma - 1|^2 d\mathbb{P} = 2 \sum_{k=1}^n (t_k - t_{k-1})^2 \to 0,$$

as $|\sigma| \to 0$. $\square$

We can now prove,

Proposition 3.8 *Let B be a Brownian motion in a probability space $(\Omega, \mathscr{F}, \mathbb{P})$ and let $T > 0$. Then, for almost all $\omega \in \Omega$, the total variation of $B(\cdot)(\omega)$ in $[0, T]$ is infinite.*

Proof. Set

$$\Gamma_1 = \{\omega \in \Omega : t \to B(t)(\omega) \text{ is continuous}\}.$$

We know by Theorem 3.1 that $\mathbb{P}(\Gamma_1) = 1$. Moreover, by Lemma 3.7 it follows that there exists a sequence (σ_n) of decompositions of $[0, T]$ and a set $\Gamma_2 \subset \mathscr{F}$ of probability 1 such that

$$\lim_{n \to \infty} J_{\sigma_n}(\omega) = T \quad \text{for all } \omega \in \Gamma_2.$$

By Exercise 3.6 we know that $V_T(B(\cdot)(\omega)) = +\infty$ for all $\omega \in \Gamma_1 \cap \Gamma_2$. Since $\mathbb{P}(\Gamma_1 \cap \Gamma_2) = 1$ the conclusion follows. $\square$

3.3 Wiener integral

Let B be a Brownian motion in a probability space $(\Omega, \mathscr{F}, \mathbb{P})$ and let $T > 0$. We want to define the integral

$$I(f) = \int_0^T f(s) dB(s),$$

for any $f \in H = L^2(0, T)$. We notice that we cannot give a meaning to the formula

$$I(f)(\omega) = \int_0^T f(s) dB(s)(\omega), \quad \omega \in \Omega,$$

for almost all $\omega \in \Omega$ because, as we have seen in the previous section, $B(\cdot)(\omega)$ has no finite total variation for almost all $\omega \in \Omega$. Thus, we shall define $I(f)$ as an element of $L^2(\Omega, \mathscr{F}, \mathbb{P})$.

Let us first define the integral on the subspace $S(0, T)$ of step functions, that is of all functions having the following form:

$$f = \sum_{j=1}^n a_j \mathbf{1}_{[t_{j-1}, t_j)}, \tag{3.14}$$

where $n \in \mathbb{N}$, $a_1, \ldots, a_n \in \mathbb{R}$, and $0 = t_0 \leq t_1 < \cdots < t_n = T$. If f is given by (3.14) we set

$$I(f) = \sum_{j=1}^n f(t_{j-1})(B(t_j) - B(t_{j-1})).$$

Proposition 3.9 *The mapping*

$$I : S(0,T) \subset L^2(0,T) \to L^2(\Omega, \mathscr{F}, \mathbb{P}), \; f \mapsto I(f),$$

can be uniquely extended to an isometry of H into $L^2(\Omega, \mathscr{F}, \mathbb{P})$. *Moreover*

$$\int_\Omega I(f)d\mathbb{P} = 0, \tag{3.15}$$

and

$$\int_\Omega |I(f)|^2 \, d\mathbb{P} = |f|^2. \tag{3.16}$$

Proof. Let $f \in S(0,T)$ of the form (3.14). Then we have

$$\int_\Omega I(f)d\mathbb{P} = \sum_{j=1}^n a_j \int_\Omega (B(t_j) - B(t_{j-1}))d\mathbb{P} = 0,$$

so that (3.16) follows. Moreover

$$\int_\Omega |I(f)|^2 \, d\mathbb{P} = \int_\Omega \sum_{j=1}^n |a_j|^2 [B(t_j) - B(t_{j-1})]^2 \, d\mathbb{P}$$

$$+ 2 \int_\Omega \sum_{j<k=1}^n f(t_{j-1})f(t_{k-1})[B(t_j) - B(t_{j-1})][B(t_k) - B(t_{k-1})] \, d\mathbb{P}.$$

$$\tag{3.17}$$

Since $B(t_j) - B(t_{j-1})$ and $B(t_k) - B(t_{k-1})$ are independent, it follows that

$$\int_\Omega |I_\sigma|^2 d\mathbb{P} = \sum_{j=1}^n |a_j|^2 (t_j - t_{j-1}) = |f|^2,$$

which yields (3.17). So, we have proved that I is an isometry. Since $S(0,T)$ is dense in $L^2(0,T)$, we can extend I to the whole $L^2(0,T)$ and the extension still fulfills (3.15) and (3.16). $\square$

The element $\int_0^T f(s)dB(s)$ of $L^2(\Omega, \mathscr{F}, \mathbb{P})$, is called the *Wiener integral* of f in $[0,T]$.

Exercise 3.10 Let f, g $L^2(0,T)$. Prove that

$$\mathbb{E}\left(\int_0^T f(s)dB(s) \int_0^T g(s)dB(s) \right) = \int_0^T f(s)g(s)ds. \tag{3.18}$$

We define in an obvious way the Wiener integral $\int_a^b f(s)dB(s)$ in any interval $[a,b] \subset \mathbb{R}$. It is easy to see that, for any $a, b, c \in [0,T]$ we have

$$\int_a^c f(s)dB(s) = \int_a^b f(s)dB(s) + \int_b^c f(s)dB(s).$$

Proposition 3.11 *Let $f \in L^2(0,T)$. Then $I(f)$ is a real Gaussian random variable with mean 0 and covariance $\int_0^T |f(s)|^2 ds$.*

Proof. Let (f_n) be a sequence in $S(0,T)$ convergent to f in $L^2(0,T)$. We know that $I(f_n) \to I(f)$ in $L^2(\Omega, \mathscr{F}, \mathbb{P})$ as $n \to \infty$. On the other hand, we have (recall Exercise 1.21),

$$ I(f_n)_\# \mathbb{P} = N_{\sum_{j=1}^n f^2(t_{j-1})(t_j - t_{j-1})}. $$

Now the conclusion follows from Proposition 1.16. $\square$

We note that if $f \in C^1([0,T])$ it is possible to express the Wiener integral $\int_0^T f(s) dB(s)$ in terms of a Riemann integral as the following integration by parts formula shows.

Proposition 3.12 *If $f \in C^1([0,T])$ we have*

$$ \int_0^T f(s) dB(s) = f(T)B(T) - \int_0^T f'(s)B(s) ds \quad \text{in } L^2(\Omega, \mathscr{F}, \mathbb{P}). \quad (3.19) $$

Proof. Let $\sigma = \{0 = t_0 < t_1 < \cdots < t_n = T\} \in \Sigma(0,T)$. Then we have

$$ I_\sigma(f) = \sum_{k=1}^n f(t_{k-1})(B(t_k) - B(t_{k-1})) $$

$$ = \sum_{k=1}^n (f(t_k)B(t_k) - f(t_{k-1})B(t_{k-1})) $$

$$ - \sum_{k=1}^n (f(t_k) - f(t_{k-1}))B(t_k) $$

$$ = f(T)B(T) - \sum_{k=1}^n (f(t_k) - f(t_{k-1}))B(t_k) $$

$$ = f(T)B(T) - \sum_{k=1}^n f'(\alpha_k)B(t_k)(t_k - t_{k-1}), $$

where α_k is a suitable number lying in the interval $[t_{k-1}, t_k], k = 1, \ldots, n$. It follows that

$$ \lim_{|\sigma| \to 0} I_\sigma(f) = f(T)B(T) - \int_0^T f'(s)B(s) ds \quad \text{in } L^2(\Omega, \mathscr{F}, \mathbb{P}). $$

$\square$

Remark 3.13 Let $B(t) = W_{1_{[0,t]}}$ be the Brownian motion in $(H, \mathscr{B}(H), \mu)$, where H is a Hilbert space and μ a Gaussian measure N_Q with Ker $Q = \{0\}$ introduced in section 3.1. Then there is a simple interpretation of the Wiener integral of a function $f \in L^2(0,T)$. In fact if $\sigma \in \Sigma$ we have

$$I_\sigma = \sum_{k=1}^n f(t_{k-1}) W_{1_{(t_{k-1},t_k]}} = W_{\sum_{k=1}^n f(t_{k-1}) 1_{(t_{k-1},t_k]}}.$$

Since

$$\lim_{|\sigma| \to 0} \sum_{k=1}^n f(t_{k-1}) 1_{(t_{k-1},t_k]} = f \quad \text{in } L^2(0,T),$$

it follows that

$$I(f) = \int_0^T f(s) dB(s) = W_f.$$

Consequently, we recover the result, proved in Proposition 3.11 that $I(f)$ is a real Gaussian random variable with law $N_{|f|^2}$.

3.4 Law of the Brownian motion in $L^2(0,T)$

We use here notations of section 3.1. In particular, we consider the probability space $(H, \mathscr{B}(H), \mu)$, where $H = L^2(0, +\infty)$ and $\mu = N_Q$, where Q is any operator in $L_1^+(H)$ such that Ker $Q = \{0\}$. Let $B(t) = W_{1_{[0,t]}}$, $t \geq 0$. As we have seen, a version of B is real Brownian motion. Let us consider the mapping

$$B : H \to L^2(0,T), \quad x \mapsto B(\cdot)(x).$$

Notice that $B \in L^2(H, \mu; L^2(0,T))$ since

$$\int_H \left(\int_0^T |B(t)(x)|^2 dt \right) \mu(dx) = \int_0^T dt \int_H |W_{1_{[0,t]}}|^2 \mu(dx)$$

$$= \int_0^T t\, dt = \frac{T^2}{2}.$$

Let us denote by $\mathbb{P} := B_\# \mu$ the law of B, it is a probability measure on $(L^2(0,T), \mathscr{B}(L^2(0,T)))$ which extends in an obvious way the Wiener measure (which was defined in $C_0([0,T])$).

Proposition 3.14 $\mathbb{P}$ *is a nondegenerate Gaussian measure on $L^2(0,T)$ with mean 0 and covariance operator $\mathscr{R}_T$ defined by*

$$\mathscr{R}_T h(t) = \int_0^T \min\{t, s\} h(s) ds, \quad h \in L^2(0,T), \ t \in [0,T]. \qquad (3.20)$$

Proof. Define for any $n \in \mathbb{N}$

$$B_n(t) = W_{P_n \mathbf{1}_{[0,t]}} = \langle x, Q_n^{-1/2} \mathbf{1}_{[0,t]} \rangle = \sum_{k=1}^{n} \lambda_k^{-1/2} x_k \int_0^t e_k(s) ds, \quad t \geq 0$$

and consider the mapping

$$B_n : H \to L^2(0, T), \quad x \mapsto B_n(\cdot)(x).$$

B_n is a linear bounded operator from H into $L^2(0, T)$. Therefore its law is Gaussian (Proposition 1.18). We claim that

$$B_n \to B \quad \text{in } L^2(H, \mu; L^2(0, T)), \tag{3.21}$$

so that the law $\mathbb{P}$ of B is also Gaussian (Proposition 1.16). We have in fact, using the Fubini theorem,

$$\int_H |B(x) - B_n(x)|_{L^2(0,T)}^2 \, \mu(dx)$$

$$= \int_H \left[\int_0^T |W_{\mathbf{1}_{[0,t]}}(x) - W_{P_n \mathbf{1}_{[0,t]}}(x)|^2 dt \right] \mu(dx)$$

$$= \int_H \left[\int_0^T |W_{(1-P_n)\mathbf{1}_{[0,t]}}(x)|^2 dt \right] \mu(dx)$$

$$= \int_0^T \left[\int_H |W_{(1-P_n)\mathbf{1}_{[0,t]}}(x)|^2 \mu(dx) \right] dt$$

$$= \int_0^T |(1 - P_n)\mathbf{1}_{[0,t]}|^2 dt.$$

So, (3.21) follows from the dominated convergence theorem.

Clearly the mean of $\mathbb{P}$ is 0; let us compute its covariance $\mathscr{R}_T$. We have, using (1.17),

$$\langle \mathscr{R}_T h, h \rangle_{L^2(0,T)} = \int_H |\langle Bx, h \rangle_H|^2 \mu(dx)$$

$$= \int_H \mu(dx) \int_0^T W_{\mathbf{1}_{[0,t]}}(x) h(t) dt \int_0^T W_{\mathbf{1}_{[0,s]}}(x) h(s) ds$$

$$= \int_0^T \int_0^T dt \, ds \, h(t) h(s) \int_H W_{\mathbf{1}_{[0,t]}}(x) W_{\mathbf{1}_{[0,s]}}(x) \mu(dx)$$

$$= \int_0^T \int_0^T \min\{t, s\} h(t) h(s) dt ds$$

and the conclusion follows.

It remains to show that the law of $\mathbb{P}$ is nondegenerate. Assume that $h \in L^2(0,T)$ is such that $\mathscr{R}_T h(t) = 0$ in $L^2(L^2(0,T), \mathscr{B}(L^2(0,T)), \mathbb{P})$. Then we have

$$\int_0^T \min\{t,s\}h(s)ds = \int_0^t sh(s)ds + t\int_t^T h(s)ds = 0, \quad t \text{ a.e. in } [0,T].$$

Differentiating this identity with respect to t yields

$$\int_t^T h(s)ds = 0, \quad t\text{-a.e. in } [0,T],$$

which implies h identically equal to 0. The proof is complete. $\square$

Exercise 3.15 Let $A_T = \mathscr{R}_T^{-1}$. Show that

$$\begin{cases} A_T h(t) = -h''(t), \ \forall\, h \in D(A_T) \\ \\ D(A) = \{h \in H^2(0,T) : h(0) = h'(T) = 0\}, \end{cases}$$

where $H^2(0,T)$ denotes the usual Sobolev space.

3.4.1 Brownian bridge

Let consider the stochastic process in $[0,T]$

$$\beta(t) : = B(t) - \frac{t}{T}B(T), \quad t \in [0,T].$$

β is called the *Brownian bridge* in $[0,T]$. Clearly β is a continuous stochastic process.

Exercise 3.16 Prove that the law of $\beta(\cdot)$ in $L^2(0,T)$ is a Gaussian measure $N_{\mathscr{S}_T}$, where $\mathscr{S}_T$ is defined by

$$\mathscr{S}_T h(t) = \int_0^T K(t,s)h(s)ds, \ h \in L^2(0,T), \quad t \in [0,1], \qquad (3.22)$$

and

$$K(t,s) = \begin{cases} s(T-t), & \text{if } 0 \le s \le t, \\ t(T-s), & \text{if } t \le s \le T. \end{cases}$$

Moreover setting $B_T = \mathscr{S}_T^{-1}$, show that

$$\begin{cases} B_T h(t) = -h''(t), & h \in D(B_T) \\ \\ D(B_T) = H^2(0,T) \cap H_0^1(0,T). \end{cases}$$

3.5 Multidimensional Brownian motions

Let $X_1, \ldots, X_n$ be real stochastic processes in a probability space $(\Omega, \mathscr{F}, \mathbb{P})$. They are said to be *independent* if for arbitrary $t_j^i \geq 0$, $i, j = 1, \ldots, n$, the random variables with values in $\mathbb{R}^n$

$$(X_i(t_1^i), \ldots, X_i(t_n^i)), \quad i = 1, \ldots, n,$$

are independent.

A *Brownian motion* in $\mathbb{R}^n$ is a stochastic process

$$B = B(t) = (B_1(t), \ldots, B_n(t)), \quad t \geq 0,$$

taking values in $\mathbb{R}^n$ such that $B_1, \ldots, B_n$ are mutually independent real Brownian motions.

Let us construct a Brownian motion in $\mathbb{R}^n$. Denote by $(e_1, \ldots, e_n)$ an orthonormal basis in $\mathbb{R}^n$. Then set $H = L^2(0, +\infty; \mathbb{R}^n)$, $\mathscr{F} = \mathscr{B}(H)$, and $\mathbb{P} = N_Q$, where Q is any operator in $L_1^+(H)$ such that Ker $Q = \{0\}$.

The following result can be proved as Theorem 3.1. The simple proof is left to the reader.

Theorem 3.17 *Let* $B_i(t) = W_{\mathbf{1}_{[0,t]}e_i}$, $t \geq 0$. *Then* $B(t) = (B_1(t), \ldots, B_n(t))$, *is a Brownian motion in* $\mathbb{R}^n$.

Remark 3.18 Let B be a Brownian motion in $\mathbb{R}^n$. Then the following properties are easy to check.

(i) For all $t > s > 0$, $B(t) - B(s)$ is a Gaussian random variable with law $N_{(t-s)I_n}$, where I_n is the identity operator in $\mathbb{R}^n$.
(ii) For all $t, s > 0$, $\mathbb{E}(B_i(t)B_j(s)) = 0$ if $i \neq j$, $i, j = 1, \ldots, n$.
(iii) We have

$$\mathbb{E}\left[|B(t) - B(s)|^2\right] = n(t - s). \tag{3.23}$$

Let us check (iii). We have

$$\mathbb{E}\left[|B(t) - B(s)|^2\right] = \sum_{k=1}^{n} \mathbb{E}\left[|B_k(t) - B_k(s)|^2\right] = n(t - s).$$

Exercise 3.19 Prove that for $0 \leq s < t$ we have

$$\mathbb{E}\left[|B(t) - B(s)|^4\right] = (2n + n^2)(t - s)^2. \tag{3.24}$$

Now we define the Wiener integral

$$F(T) = \int_0^T G(t) dB(t),$$

for a function $G \in C([0,T]; L(\mathbb{R}^n))$. Set

$$G_{h,k}(t) = \langle G(t)e_k, e_h \rangle, \quad t \in [0,1], \ h, k = 0, 1, \ldots, n,$$

and

$$F(T)_h = \sum_{k=1}^{n} \int_0^1 G_{h,k}(t) dB_k(t), \quad h = 1, \ldots, n. \tag{3.25}$$

Proposition 3.20 *Let* $G \in C([0,T]; L(\mathbb{R}^n))$, *and let*

$$F(T) = (F(T)_1, \ldots, F(T)_n)$$

be defined by (3.25). Then we have

$$\mathbb{E}(F(T)) = 0, \tag{3.26}$$

and

$$\mathbb{E}(|F(T)|^2) = \int_0^T \mathrm{Tr}\ [G(t)G^*(t)]dt. \tag{3.27}$$

Proof. (3.26) is obvious, let us prove (3.27). We have in fact

$$\mathbb{E}(|I|^2) = \sum_{h=1}^{n} \mathbb{E}(|I_h|^2)$$

$$= \mathbb{E}\left(\sum_{h,k,p=1}^{n} \int_0^T G_{h,k}(t) dB_k(t) \int_0^1 G_{h,p}(t) dB_p(t) \right)$$

$$= \sum_{h,k=1}^{n} \int_0^T G_{h,k}^2(t) dt = \int_0^T \mathrm{Tr}\ [G(t)G^*(t)]dt.$$

□

Exercise 3.21 Let $X(t) = \int_0^t G(s)dB(s)$. Prove that $X(t)$ is a Gaussian random variable with law N_{Q_t} where

$$Q_t = \int_0^t \mathrm{Tr}\ [G(s)G^*(s)]ds.$$

Stochastic perturbations of a dynamical system

4.1 Introduction

In this chapter we are given a Brownian motion $B(t)$, $t \geq 0$, in a probability space $(\Omega, \mathscr{F}, \mathbb{P})$ with values in $\mathbb{R}^n$. Without any loss of generality we can assume that $B(\cdot)(\omega)$ is continuous for all $\omega \in \Omega$.

Let us consider a dynamical system in $\mathbb{R}^n$ governed by the ordinary differential equation

$$\begin{cases} Z'(t) = b(Z(t)), & t \geq 0, \\ \\ Z(0) = x \in \mathbb{R}^n, \end{cases} \tag{4.1}$$

where $b \colon \mathbb{R}^n \to \mathbb{R}^n$ is Lipschitz continuous, that is there exists $M > 0$ such that

$$|b(x) - b(y)| \leq M|x - y|, \quad x, y \in \mathbb{R}^n.$$

It is well known that equation (4.1) is equivalent to the integral equation

$$Z(t) = x + \int_0^t b(Z(s))ds, \quad t \geq 0 \tag{4.2}$$

and that equation (4.2) has a unique solution $Z(\cdot, x) \in C^1([0, +\infty); \mathbb{R}^n)$.

To take into account random perturbations one is led to consider the following *stochastic differential* equation,

$$X(t) = x + \int_0^t b(X(s))ds + \sqrt{C}\, B(t), \quad t \geq 0, \tag{4.3}$$

where C is a symmetric and non-negative linear operator $\mathbb{R}^n \to \mathbb{R}^n$. The unknown $X(\cdot)$ of (4.3) is a stochastic process on $(\Omega, \mathscr{F}, \mathbb{P})$.

Usually, equation (4.3) is formally written as

$$\begin{cases} dX(t) = b(X(t))dt + \sqrt{C}\, dB(t), \\ X(0) = x. \end{cases} \tag{4.4}$$

Setting

$$X(t,\omega) = X(t)(\omega), \quad B(t,\omega) = B(t)(\omega), \quad t \geq 0, \ \omega \in \Omega,$$

equation (4.3) can be regarded as a family of deterministic integral equations indexed by ω,

$$X(t,\omega) = x + \int_0^t b(X(s,\omega))ds + \sqrt{C}\, B(t,\omega), \quad t \geq 0. \tag{4.5}$$

Consequently, to solve the integral equation (4.5), it is enough to solve for any $x \in \mathbb{R}^n$ and $f \in C([0,T];\mathbb{R}^n)$ [1] the deterministic integral equation

$$u(t) = x + \int_0^t b(u(s))ds + f(t), \quad t \in [0,T], \tag{4.6}$$

whose solution we shall denote by $u(\cdot, f)$, and then to set

$$X(\cdot, x)(\omega) = u(x, \sqrt{C}\, B(\cdot)(\omega)), \quad \omega \in \Omega.$$

Remark 4.1 When $f \subset C^1([0,T];H)$, equation (4.6) coincides with the Cauchy problem

$$\begin{cases} u'(t) = b(u(t)) + f'(t), \\ u(0) = x + f(0). \end{cases} \tag{4.7}$$

Equation (4.6) can be easily solved by the classical method of successive approximations. That is, setting

$$u_0(t) = x, \quad u_{n+1}(t) = x + \int_0^t b(u_n(s))ds + f(t), \quad n \in \mathbb{N}, \ t \geq 0, \tag{4.8}$$

the following result holds,

---

[1] $C([0,T];\mathbb{R}^n)$ is the space of all continuous functions $f \colon [0,T] \to \mathbb{R}^n$ endowed with the norm $\|f\|_0 = \sup_{t \in [0,T]} |f(t)|$. $C^1([0,T];\mathbb{R}^n)$ is the subspace of $C([0,T];\mathbb{R}^n)$ of all continuously differentiable functions. We set $\|f\|_1 = \|f\|_0 + \|Df\|_0$.

Lemma 4.2 *Let $x \in \mathbb{R}^n$, $T > 0$, $f \in C([0,T];\mathbb{R}^n)$. Then there exists a unique $u \in C([0,T];\mathbb{R}^n)$ fulfilling equation (4.6) and we have*

$$u = \lim_{n \to \infty} u_n \quad \text{in } C([0,T];\mathbb{R}^n). \tag{4.9}$$

Moreover, denoting by γ_T the mapping

$$\gamma_T \colon \mathbb{R}^n \times C([0,T];\mathbb{R}^n) \to C([0,T];\mathbb{R}^n), \quad (x,f) \mapsto \gamma_T(x,f) = u, \tag{4.10}$$

we have

$$\|\gamma_T(x,f) - \gamma_T(x_1,f_1)\|_0 \le e^{TM}|x - x_1| + \int_0^T e^{sM} ds \, \|f - f_1\|_0, \tag{4.11}$$

for any $x, x_1 \in \mathbb{R}^n$, and any $f, f_1 \in C([0,T];\mathbb{R}^n)$.

Finally, if in addition b is of class C^k for some $k \in \mathbb{N}$, then γ_T is of class C^k.

Proof. We only sketch the proof (which is very similar to that of existence and uniqueness for the Cauchy problem in $\mathbb{R}^n$). We have

$$|u_1(t) - u_0(t)| \le |b(x)|T + \|f\|_0, \quad t \in [0,T],$$

and, by recurrence on n,

$$|u_{n+1}(t) - u_n(t)| \le (|b(x)|T + \|f\|_0) \frac{M^n T^n}{n!}, \quad n \in \mathbb{N}, \, t \in [0,T],$$

and so (4.9) follows easily by a classical argument.

Let us prove (4.11). Let $x, x_1 \in \mathbb{R}^n$, $f, f_1 \in C([0,T];\mathbb{R}^n)$ and let $u = \gamma_T(x,f)$, $u_1 = \gamma_T(x_1,f_1)$. Then we have

$$u(t) - u_1(t) = x - x_1 + \int_0^t [b(u(s)) - b(u_1(s))]ds + (f(t) - f_1(t)), \quad t \in [0,T].$$

It follows that for $t \in [0,T]$,

$$|u(t) - u_1(t)| \le |x - x_1| + M \int_0^t |u(s) - u_1(s)|ds + |f(t) - f_1(t)|.$$

Consequently, by the Gronwall lemma we have

$$|u(t) - u_1(t)| \le e^{tM}|x - x_1| + \int_0^t e^{(t-s)M}|f(s) - f_1(s)|ds, \quad t \in [0,T], \tag{4.12}$$

which yields (4.11).

Finally the last statement is standard, its proof it left as an exercise to the reader. $\square$

Now we come back to the stochastic differential equation (4.3).

Proposition 4.3 *Let $\eta \in L^2(\Omega, \mathscr{F}, \mathbb{P}; \mathbb{R}^n)$. Then the following statements hold.*

(i) There exists a unique continuous stochastic process $X(\cdot, \eta)$ in $[0, +\infty)$ solution of the integral equation

$$X(t, \eta) = \eta + \int_0^t b(X(s, \eta))ds + \sqrt{C}\, B(t). \tag{4.13}$$

$X(\cdot, \eta)$ *is given by*

$$X(\cdot, \eta(\omega)) = \gamma_T(\eta(\omega), \sqrt{C}\, B(\cdot)(\omega)), \quad \omega \in \Omega, \tag{4.14}$$

where γ_T is defined by (4.10).
(ii) For any $T > 0$ we have

$$X(\cdot, \eta) = \lim_{n \to \infty} X_n(\cdot, \eta) \quad in \ C([0, T]; L^2(\Omega, \mathscr{F}, \mathbb{P}; \mathbb{R}^n)), \tag{4.15}$$

where X_n, $n \in \mathbb{N}$, is defined by recurrence as

$$X_0(t, \eta) = \eta, \ X_{n+1}(t, \eta) = \eta + \int_0^t b(X_n(s, \eta))ds + \sqrt{C}\, B(t), \ t \in [0, T].$$

(iii) If $\eta = x$ is constant, the law of $X(\cdot, x)$ is independent of the choice of the particular Brownian motion B.

Proof. (i) and (ii) are immediate consequences of Lemma 4.2. Finally, since

$$X(\cdot, x) = \gamma_T(x, \sqrt{C}\, B(\cdot)), \tag{4.16}$$

the law of $X(\cdot, x)$, is determined by that of $B(\cdot)$ which does not depend on the choice of B (recall Proposition 3.14). $\square$

Exercise 4.4 Show that if $\eta = x$ is constant and $t, h > 0$, the random variables $X(t, x)$ and $B(t + h) - B(t)$, are independent.
Hint. Check by recurrence that $X_n(t, x)$ and $B(t + h) - B(t)$ are independent.

Remark 4.5 It is useful to study problems with a general initial time $s \in \mathbb{R}$,
$$\begin{cases} Z'(t, x) = b(Z(t, x)), & t \geq s, \\ Z(s, x) = x \in H. \end{cases}$$

This problem is equivalent to the integral equation

$$Z(t,x) = x + \int_s^t b(Z(u,x))du, \quad t \geq 0.$$

Similarly to the stochastic case we shall consider the equation

$$X(t,s,\eta) = \eta + \int_s^t b(X(\rho,\eta))d\rho + \sqrt{C}\,(B(t) - B(s)), \quad t \geq s,$$

$$(4.17)$$

where $\eta \in L^2(\Omega, \mathscr{F}, \mathbb{P}; \mathbb{R}^n)$. Then, proceding as before, we can see that the equation (4.17) has a unique solution $X(\cdot, s, \eta)$. Moreover we have clearly

$$X(t,\eta) = X(t,0,\eta), \quad t \geq 0, \quad \eta \in L^2(\Omega, \mathscr{F}, \mathbb{P}; \mathbb{R}^n). \quad (4.18)$$

Exercise 4.6 Prove that if $0 \leq s < \sigma < t$, and $\eta \in L^2(\Omega, \mathscr{F}, \mathbb{P}; \mathbb{R}^n)$, we have

$$X(t, \sigma, X(\sigma, s, \eta)) = X(t, s, \eta). \quad (4.19)$$

Hint: Use the uniqueness result for problem (4.17).

Proposition 4.7 *Let $x \in H$, $t, s, h > 0$, $t > s$. Then the random variables $X(t+h, s+h, x)$ and $X(t, s, x)$ have the same law.*

Proof. Write

$$X(t,s,x) = x + \int_s^t b(X(u,s,x))du + \sqrt{C}\,(B(t) - B(s)) \quad (4.20)$$

and

$$X(t+h, s+h, x) = x + \int_{s+h}^{t+h} b(X(u, s+h, x))du \\ + \sqrt{C}\,(B(t+h) - B(s+h)). \quad (4.21)$$

Setting $u = v + h$ and $B_1(t) = B(t+h) - B(h)$, we can write (4.21) as

$$X(t+h, s+h, x) = x + \int_s^t b(X(v+h, s+h, x))dv \\ + \sqrt{C}\,(B_1(t) - B_1(s)). \quad (4.22)$$

Recall that $B_1(t)$ is a Brownian motion by the invariance by translation (see Exercise 3.5). This shows that the process $X(\cdot + h, s + h, x)$ fulfills (4.20) with the Brownian motion $B(t)$ replaced by $B_1(t)$. Thus the law of $X(t+h, s+h, x)$ coincides with that of $X(t, s, x)$ thanks to Proposition 4.3(iii). $\square$

Exercise 4.8 Show that if $\eta \in L^2(\Omega, \mathscr{F}, \mathbb{P}; \mathbb{R}^n)$, then the laws of $X(t, s, \eta)$ and $X(t + h, s + h, \eta)$ are different in general.
Hint: Take $\eta = B(s)$.

4.2 The Ornstein–Uhlenbeck process

We assume here that $b(x) = Ax$, where $A \in L(\mathbb{R}^n)$. In this case we can solve explicitly equations (4.3) and (4.6).

Lemma 4.9 Let $A \in L(\mathbb{R}^n)$, $x \in H$ and $f \in C([0, T]; \mathbb{R}^n)$. Then the solution to the equation

$$u(t) = x + \int_0^t Au(s)ds + f(t), \quad t \geq 0, \tag{4.23}$$

is given by

$$u(t) = e^{tA}x + f(t) + \int_0^t Ae^{(t-s)A}f(s)ds, \quad t \geq 0. \tag{4.24}$$

Proof. Recall that the solution of (4.23) is given by $u = \gamma_T(x, f)$ where γ_T is defined by (4.10), and that γ_T is continuous. Since $C^1([0, T]; \mathbb{R}^n)$ is dense in $C([0, T]; \mathbb{R}^n)$, it is enough to prove (4.24) when $f \in C^1([0, T]; \mathbb{R}^n)$. In this case (4.23) is equivalent to the initial value problem

$$\begin{cases} u'(t) = Au(t) + f'(t), \\ u(0) = x + f(0), \end{cases}$$

whose solution is given by the variation of constants formula,

$$u(t) = e^{tA}(x + f(0)) + \int_0^t e^{(t-s)A}f'(s)ds.$$

Now, integrating by parts we obtain (4.24). $\square$

Proposition 4.10 Let $A \in L(\mathbb{R}^n)$, $x \in \mathbb{R}^n$. Then the solution to the stochastic differential equation

$$dX = AX dt + \sqrt{C}\, dB(t), \quad X(0) = x, \tag{4.25}$$

is given by

$$X(t, x) = e^{tA}x + \int_0^t e^{(t-s)A}\sqrt{C}\, dB(s). \tag{4.26}$$

$X(\cdot, x)$ is called an *Ornstein–Uhlenbeck* process.

Proof. Taking into account (4.24) we have

$$X(t, x) = e^{tA}x + \sqrt{C}\,B(t) + \int_0^t Ae^{(t-s)A}\sqrt{C}\,B(s)ds.$$

Now the conclusion follows from the integration by parts formula (3.19).
□

Exercise 4.11 Prove that the law of $X(t, x)$ is given by

$$X(t, x)_{\#}\mathbb{P} = N_{e^{tA}x, Q_t}, \tag{4.27}$$

where

$$Q_t = \int_0^t e^{sA}Ce^{sA^*}ds. \tag{4.28}$$

Hint. Use Exercise 3.21.

4.3 The transition semigroup in the deterministic case

We are here concerned with problem (4.1) under the assumption that $b\colon \mathbb{R}^n \to \mathbb{R}^n$ is Lipschitz continuous and of class C^1. We know that for any $x \in \mathbb{R}^n$ problem (4.1) has a unique solution $Z(\cdot, x) \in C^1([0, +\infty); H)$ and

$$Z(t + s, x) = Z(t, Z(s, x)). \tag{4.29}$$

Moreover $Z(\cdot, x)$ is differentiable in x as the following proposition shows.

Proposition 4.12 *For any $t \geq 0$, $Z(t, x)$ is differentiable on x and we have*

$$\langle Z_x(t, x), h \rangle = \eta^h(t, x), \quad x, h \in H, \ t \geq 0, \tag{4.30}$$

where $\eta^h(t, x)$ is the solution to the initial value problem

$$\begin{cases} \dfrac{d}{dt}\,\eta^h(t, x) = b_x(Z(t, x)) \cdot \eta^h(t, x), \\[2mm] \eta^h(0, x) \quad = h. \end{cases} \tag{4.31}$$

Moreover,

$$b(Z(t, x)) = Z_x(t, x)b(x), \quad x \in H, \ t \geq 0. \tag{4.32}$$

Proof. This proposition is well known. We only prove (4.32) for the reader's convenience. Differentiating (4.29) with respect to s yields

$$Z_t(t + s, x) = Z_x(t, Z(s, x)) \cdot Z_t(s, x),$$

which is equivalent to

$$b(Z(t + s, x)) = Z_x(t, Z(s, x)) \cdot b(Z(s, x)).$$

Setting $s = 0$, the identity (4.32) follows. $\square$

Now we associate to the dynamical system (4.1) (which is obviously non-linear in general) a semigroup of *linear* operators defined on the space $C_b(\mathbb{R}^n)$ [2] setting,

$$P_t\varphi(x) = \varphi(Z(t, x)), \quad x \in H, \ t \geq 0.$$

P_t is called the *transition semigroup* related to the dynamical system described by (4.1).

In the applications to physics a function $\varphi \in C_b(\mathbb{R}^n)$ is often interpreted as an "observable". Then $P_t\varphi$ describes the evolution in time of the observable. The asymptotic behaviour of $P_t\varphi$ gives important information on the dynamical system (4.1) as: invariant measures, ergodicity, mixing etc.; concepts that we shall introduce in the next chapter.

We notice that from (4.29) it follows immediately that the semigroup law holds, namely

$$P_{t+s} = P_t P_s, \quad t, s \geq 0. \tag{4.33}$$

The transition semigroup P_t is related to the following partial differential equation of the first order, [3]

$$\begin{cases} v_t(t, x) = \langle b(x), D_x v(t, x) \rangle, \\ v(0, x) = \varphi(x), \end{cases} \tag{4.34}$$

where $\varphi \in C_b^1(\mathbb{R}^n)$.

By a *strict* solution of (4.34) we mean a function $v : [0, T] \times \mathbb{R}^n \to \mathbb{R}^n$ of class C^1 such that (4.34) holds.

---

[2] $C_b(\mathbb{R}^n)$ is the Banach space of all uniformly continuous and bounded mappings $\varphi \colon \mathbb{R}^n \to \mathbb{R}$, endowed with the norm $\|\varphi\|_0 = \sup_{x \in \mathbb{R}^n} |\varphi(x)|$. For any $k \in \mathbb{N}$, $C_b^k(\mathbb{R}^n)$ is the subspace of $C_b(\mathbb{R}^n)$ of all functions which are continuous and bounded together with their derivatives of order less than or equal to k. We set $\|\varphi\|_k = \|\varphi\|_0 + \sum_{j=1}^k \sup_{x \in H} |D_x^j \varphi(x)|$.

[3] If $\varphi \in C_b^1(\mathbb{R}^n)$ and $x \in \mathbb{R}^n$ we shall identify $D_x\varphi(x)$ with the unique element h of $\mathbb{R}^n$ such that $D_x\varphi(x)y = \langle h, y \rangle$, $\forall y \in \mathbb{R}^n$. If $\varphi \in C_b^2(\mathbb{R}^n)$ and $x \in \mathbb{R}^n$ we shall identify $D_x^2\varphi(x)$ with the unique element T of $L(\mathbb{R}^n)$ such that $D_x^2\varphi(x)(y, z) = \langle Ty, z \rangle$, $\forall y, z \in \mathbb{R}^n$.

Theorem 4.13 *Let $\varphi \in C_b^1(\mathbb{R}^n)$. Then problem (4.34) has a unique strict solution v given by*

$$v(t, x) = \varphi(Z(t, x)) = P_t\varphi(x), \quad t \geq 0, \ x \in \mathbb{R}^n. \qquad (4.35)$$

Proof. *Existence.* We shall prove that $v(t, x)$, given by (4.35), is a strict solution of (4.34). We have in fact, for $t \geq 0$, $x \in \mathbb{R}^n$,

$$v_t(t, x) = \langle D_x\varphi(Z(t, x)), Z_t(t, x)\rangle = \langle D_x\varphi(Z(t, x)), b(Z(t, x))\rangle. \quad (4.36)$$

Moreover, taking into account (4.32), we have

$$\begin{aligned}
\langle D_x v(t, x), b(x)\rangle &= \langle D_x\varphi(Z(t, x)), Z_x(t, x) \cdot b(x)\rangle \\
&= \langle D_x\varphi(Z(t, x)), b(Z(t, x))\rangle.
\end{aligned}$$

Comparing with (4.36) yields the conclusion.

Uniqueness. Let ζ be a strict solution of (4.34). Fix $t > 0$. Then for any $s \in [0, t]$ we have

$$\frac{d}{ds}\,\zeta(t - s, Z(s, x)) = -\zeta_t(t - s, Z(s, x))$$

$$+ \langle D_x\zeta(t - s, Z(s, x)), b(Z(s, x))\rangle = 0.$$

It follows that $\zeta(t - s, Z(s, x))$ is constant for $s \in [0, t]$. Setting $s = t$ and $s = 0$ we see that $\varphi(Z(t, x)) = \zeta(t, x)$. $\square$

4.4 The transition semigroup in the stochastic case

Here we want to associate to the solution $X(t, x)$ of the stochastic differential equation

$$X(t, x) = x + \int_0^t b(X(s, x))ds + \sqrt{C}\,B(t), \quad t \geq 0, \ x \in H \qquad (4.37)$$

a transition semigroup P_t. It is natural to set

$$P_t\varphi(x) = \int_\Omega \varphi(X(t, x)(\omega))\mathbb{P}(d\omega) = \mathbb{E}\left[\varphi(X(t, x))\right], \quad t \geq 0, \ \varphi \in C_b(\mathbb{R}^n).$$
$$(4.38)$$

Remark 4.14 Let $T > 0$, $H = L^2(0, T; \mathbb{R}^n)$. Then, for any $t \in [0, T]$ and any $\varphi \in C_b(\mathbb{R}^n)$ we have the following *explicit* representation formula for $P_t\varphi$,

$$P_t\varphi(x) = \int_H \varphi(\gamma_T(x, \sqrt{C}\,f)(t))N_{\mathscr{R}_T}(df), \quad x \in \mathbb{R}^n,$$

where γ_T is defined by (4.10) and $\mathscr{R}_T$ by (3.20).

In the following proposition we show that P_t acts on $C_b(\mathbb{R}^n)$. Later we shall see that P_t is a semigroup of linear bounded operators on $C_b(\mathbb{R}^n)$.

Proposition 4.15 *For any $t > 0$ and any $\varphi \in C_b(\mathbb{R}^n)$, we have $P_t\varphi \in C_b(\mathbb{R}^n)$. Moreover $P_t \in L(C_b(\mathbb{R}^n))$ and*

$$\|P_t\varphi\|_0 \leq \|\varphi\|_0. \tag{4.39}$$

Proof. Notice first that for any $\varphi \in C_b(\mathbb{R}^n)$ we have

$$|P_t\varphi(x)| \leq \mathbb{E}|\varphi(X(t,x))| \leq \|\varphi\|_0, \quad x \in \mathbb{R}^n,$$

so that (4.39) holds.

Let now $\varphi \in C_b^1(\mathbb{R}^n)$, $x, \bar{x} \in \mathbb{R}^n$. Then, recalling (4.11) we find that

$$|P_t\varphi(x) - P_t\varphi(\bar{x})| \leq \|\varphi\|_1 \mathbb{E}|X(t,x) - X(t,\bar{x})| \leq \|\varphi\|_1 e^{tM} |x - \bar{x}|.$$

Thus $P_t\varphi \in C_b(\mathbb{R}^n)$. Finally, let $\varphi \in C_b(\mathbb{R}^n)$ and $t \geq 0$. Since $C_b^1(\mathbb{R}^n)$ is dense in $C_b(\mathbb{R}^n)$, there exists a sequence $(\varphi_n) \subset C_b^1(\mathbb{R}^n)$ such that $\|\varphi - \varphi_n\|_0 \leq 1/n$ for all $n \in \mathbb{N}$. Then, since

$$\|P_t\varphi - P_t\varphi_n\|_0 \leq \|\varphi - \varphi_n\|_0 \leq \frac{1}{n},$$

we have $P_t\varphi \in C_b(\mathbb{R}^n)$. $\square$

Exercise 4.16 (i) Prove that if $\varphi \in C_b(\mathbb{R}^n)$ is non-negative, then for all $t \geq 0$ $P_t\varphi$ is non-negative as well.

(ii) Prove that $P_t(1) = 1$.

(iii) Denote by $\pi_t(x, \cdot)$ the law of $X(t,x)$, that is $\pi_t(x, \cdot) = X(t,x)_\#\mathbb{P}$. Prove that

$$P_t\varphi(x) = \int_H \varphi(y)\pi_t(x, dy), \quad \varphi \in C_b(\mathbb{R}^n). \tag{4.40}$$

Now we are going to prove that

$$\frac{d}{dt} P_t\varphi = P_t L\varphi, \quad \varphi \in C_b^2(\mathbb{R}^n), \tag{4.41}$$

and

$$\frac{d}{dt} P_t\varphi = L P_t\varphi, \quad \varphi \in C_b^2(\mathbb{R}^n), \tag{4.42}$$

where L is the following differential operator, called the *Kolmogorov* operator,

$$L\varphi(x) = \frac{1}{2}\,\mathrm{Tr}\,[CD_x^2\varphi(x)] + \langle b(x), D_x\varphi(x)\rangle, \quad \varphi \in C_b^2(\mathbb{R}^n), \ x \in \mathbb{R}^n.$$

Using (4.41) and (4.42) we will show that the parabolic equation

$$
\begin{cases}
D_t u(t, x) = \frac{1}{2}\,\mathrm{Tr}\,[CD_x^2 u(t, x)] + \langle b(x), D_x u(t, x)\rangle \\[2mm]
u(0, x) = \varphi(x), \quad x \in H,
\end{cases}
\tag{4.43}
$$

has a unique solution given by $u(t, x) = P_t\varphi(x)$, and that the *semigroup law*, called the *Chapman–Kolmogorov equation*,

$$P_{t+s} = P_t P_s, \quad t, s \geq 0, \ P_0 = 1, \tag{4.44}$$

holds.

Proposition 4.17 (Forward Itô formula) *For any* $\varphi \in C_b^2(\mathbb{R}^n)$ *we have*

$$\frac{d}{dt}\,\mathbb{E}\,[\varphi(X(t, x))] = \mathbb{E}\,[(L\varphi)(X(t, x))], \quad t \geq 0, \ x \in \mathbb{R}^n, \tag{4.45}$$

which is equivalent to (4.41).

Proof. Since C is symmetric, there exists an orthonormal basis (e_k) on $\mathbb{R}^n$ and non-negative numbers (γ_k) such that $Ce_k = \gamma_k e_k$, $k = 1, \ldots, n$. We set $x_k = \langle x, e_k\rangle$, $k = 1, \ldots, n$.

Let $h > 0$ and set $X(t+h, x) - X(t, x) = \delta_h$. By the Taylor formula we have

$$\frac{1}{h}\,[\mathbb{E}(\varphi(X(t+h, x))) - \mathbb{E}(\varphi(X(t, x)))] = I_1(h) + I_2(h) + I_3(h), \tag{4.46}$$

where

$$I_1(h) = \frac{1}{h}\,\mathbb{E}\langle D_x\varphi(X(t, x)), \delta_h\rangle$$

$$I_2(h) = \frac{1}{2h}\,\mathbb{E}\langle D_x^2\varphi(X(t, x)) \cdot \delta_h, \delta_h\rangle$$

$$I_3(h) = \frac{1}{h}\,\mathbb{E}\int_0^1 (1 - \xi)\langle[D_x^2\varphi(z(t, h, x, \xi)) - D_x^2\varphi(X(t, x))] \cdot \delta_h, \delta_h\rangle d\xi$$

and
$$z(t, h, x, \xi) = (1 - \xi)X(t, x) + \xi X(t + h, x).$$

On the other hand, we have

$$\delta_h = \int_t^{t+h} b(X(\tau, x))d\tau + \sqrt{C}\,(B(t + h) - B(t))$$

and so,

$$I_1(h) = \mathbb{E}\left\langle D_x\varphi(X(t, x)), \frac{1}{h}\int_t^{t+h} b(X(\tau, x))d\tau \right\rangle$$

$$+ \frac{1}{h}\,\mathbb{E}\langle D_x\varphi(X(t, x)), \sqrt{C}\,(B(t + h) - B(t))\rangle.$$

Since (recall Exercise 4.4), $B(t + h) - B(t)$ is independent of $D_x\varphi$ $(X(t, x))$, we have that the second integral vanishes. Thus, letting h tend to 0, we find

$$\lim_{h\to 0} I_1(h) = \mathbb{E}\left[\langle D_x\varphi(X(t, x)), b(X(t, x))\rangle\right]. \tag{4.47}$$

Concerning $I_2(h)$ we have, setting

$$F_{t,t+h} = \int_t^{t+h} b(X(\tau, x))d\tau,$$

$$I_2(h) - \frac{1}{2}\,\mathbb{E}\left\langle D_x^2\varphi(X(t, x)) \cdot F_{l,l+h}, \frac{1}{h} F_{l,l+h} \right\rangle$$

$$+ \mathbb{E}\left\langle D_x^2\varphi(X(t, x)) \cdot F_{t,t+h}, \sqrt{C}\,(B(t + h) - B(t)) \right\rangle$$

$$+ \frac{1}{2h}\,\mathbb{E}\langle D_x^2\varphi(X(t, x)) \cdot \sqrt{C}\,(B(t + h) - B(t)),$$

$$= I_{2,1}(h) + I_{2,2}(h) + I_{2,3}(h).$$

Clearly

$$\lim_{h\to 0} I_{2,1}(h) = \lim_{h\to 0} I_{2,2}(h) = 0. \tag{4.48}$$

Concerning $I_{2,3}$ we have (recall Exercise 4.4),

$$I_{2,3}(h) = \frac{1}{2h}\sum_{i=1}^n \gamma_i\,\mathbb{E}[D_{x_ix_i}^2\varphi(X(t, x))]\mathbb{E}[(B_i(t + h) - B_i(t))^2]$$

$$= \frac{1}{2}\,\mathbb{E}[\mathrm{Tr}\,(CD_x^2\varphi(X(t, x)))].$$

Therefore
$$\lim_{h\to 0} I_{2,3}(h) = \frac{1}{2}\, \mathbb{E}[\mathrm{Tr}\,(CD_x^2\varphi(X(t,x)))]. \qquad (4.49)$$

Let us show finally that
$$\lim_{h\to 0} I_3(h) = 0. \qquad (4.50)$$

For this we set
$$I_3 = \frac{1}{h}\, \mathbb{E}\int_0^1 \langle g(\xi,h)\cdot\delta_h, \delta_h\rangle d\xi,$$

where
$$g(\xi,h) = (1-\xi)\left[D_x^2\varphi((1-\xi)X(t,x) + \xi X(t+h,x)) - D_x^2\varphi(X(t,x))\right].$$

By the Hölder inequality we have
$$|I_3|^2 \le \frac{1}{h^2}\int_0^1 \mathbb{E}\|g(\xi,h)\|^2 d\xi\, \mathbb{E}\,|\delta_h|^4.$$

Thus, to prove (4.50) it is enough to show that

(i) $\lim\limits_{h\to 0}\mathbb{E}\int_0^1 \|g(\xi,h)\|^2 d\xi = 0,$

(ii) $\dfrac{1}{h^2}\,\mathbb{E}\,|\delta_h|^4 \le c$, for some $c > 0$.

(i) follows from the dominated convergence theorem since $\lim\limits_{h\to 0} g(\xi,h) = 0$, for all $h \in (0,1]$ and $\mathbb{P}$-a.s. and $\|g(\xi,h)\| \le 2\|\varphi\|_2$. Let us prove (ii). Recalling that, by (3.24),
$$\mathbb{E}\,|B(t+h) - B(t)|^4 = (2n+n^2)h^2,$$

we have
$$\frac{1}{h^2}\,\mathbb{E}\,|\delta_h|^4 \le \frac{8}{h^2}\,\mathbb{E}\left|\int_t^{t+h} b(X(s,x))ds\right|^4 + \frac{8}{h^2}\,\mathbb{E}\,|B(t+h) - B(t)|^4$$
$$\le \frac{8}{h^2}\,\mathbb{E}\left|\int_t^{t+h} b(X(s,x))ds\right|^4 + 2n + n^2,$$

and (ii) is proved.

Now the conclusion follows from (4.47)–(4.50). $\square$

Proposition 4.18 (Backward Itô formula) *Assume that b is Lipschitz continuous and of class C^2, and that $\varphi \in C_b^2(\mathbb{R}^n)$. Then we have*
$$\frac{d}{dt}\,\mathbb{E}\left[\varphi(X(t,x))\right] = L\mathbb{E}\left[\varphi(X(t,x))\right], \quad t \ge 0,\ x \in H, \qquad (4.51)$$

which is equivalent to (4.42).

Proof. Let $x \in H$, $s > 0$ be fixed. If $h > 0$ and $t > s + h$ we have, taking into account (4.19)

$$\Sigma_h := \frac{1}{h} \left[\mathbb{E}(\varphi(X(t, s + h, x))) - \mathbb{E}(\varphi(X(t, s, x))) \right]$$

$$= \frac{1}{h} \left[\mathbb{E}(\varphi(X(t, s + h, x))) - \mathbb{E}(\varphi(X(t, s + h, X(s + h, s, x)))) \right].$$

Let us consider the random variable

$$g(y) = \varphi(X(t, s + h, y)), \quad y \in H,$$

which is of class C^2 in y by Lemma 4.2, since b is of class C^2. So, we have

$$\Sigma_h = \frac{1}{h} \left[\mathbb{E}(g(x)) - \mathbb{E}(g(X(s + h, s, x))) \right].$$

Setting $\delta_h = X(s + h, s, x) - x$ and using the Taylor formula we have

$$\Sigma_h = -\frac{1}{h} \mathbb{E}\langle D_x g(x), \delta_h \rangle - \frac{1}{2h} \mathbb{E}\langle D_x^2 g(x) \cdot \delta_h, \delta_h \rangle$$

$$-\frac{1}{h} \mathbb{E} \int_0^1 (1 - \xi)\langle [D_x^2 g((1 - \xi)x + \xi X(s + h, s, x)) - D_x^2 g(x)] \cdot \delta_h, \delta_h \rangle d\xi.$$

$$= J_1 + J_2 + J_3. \tag{4.52}$$

Now, terms J_1, J_2, and J_3 can be estimated by arguing as in the proof of Proposition 4.17 and we arrive at the conclusion. $\square$

We can prove finally the result

Theorem 4.19 *Assume that b is Lipschitz continuous and of class C^2, and that $\varphi \in C_b^2(\mathbb{R}^n)$. Then problem (4.43) has a unique strict solution given by $u(t, x) = P_t\varphi(x)$, $t \geq 0$, $x \in H$. Moreover*

$$P_{t+s} = P_t P_s, \quad t, s \geq 0.$$

Proof. *Existence.* Let $u(t, x) = P_t\varphi(x)$, $t \geq 0$, $x \in H$, and set $u(t, \cdot) = u(t)$. Then by the backward Itô formula (4.51) we have

$$\frac{d}{dt} u(t) = Lu(t).$$

Uniqueness. Let u be a solution of (4.43). Then for any $t > s \geq 0$ we have by the forward Itô formula (4.45)

$$\frac{d}{ds} P_{t-s} u(s) = -P_{t-s} L u(s) + P_{t-s} \frac{d}{ds} u(s) = 0.$$

Thus the function $s \to P_{t-s} u(s)$ is constant and the conclusion follows. $\square$

Remark 4.20 Often one has to compute $\mathbb{E}\left[\varphi(X(t, x))\right]$ for some unbounded function φ. This can be done by introducing suitable approximations of φ. It is particularly important in the case when $\varphi(x) = |x|^2$. Here a direct computation is possible by the next proposition.

Proposition 4.21 *We have*

$$\frac{d}{dt} \mathbb{E}\left[|X(t, x)|^2\right] = \mathrm{Tr}[C] + 2\mathbb{E}\langle b(X(t, x)), X(t, x)\rangle, \quad t \geq 0, \ x \in H. \tag{4.53}$$

Proof. In fact, since

$$X(t + h, x) = X(t, x) + \int_t^{t+h} b(X(s, x))ds + \sqrt{C}\left(B(t + h) - B(t)\right),$$

we have

$$|X(t + h, x)|^2 - |X(t, x)|^2 = \left|\int_t^{t+h} b(X(s, x))ds\right|^2$$

$$+ |\sqrt{C}\left(B(t + h) - B(t)\right)|^2 + 2\left\langle X(t, x), \int_t^{t+h} b(X(s, x))ds\right\rangle$$

$$+ 2\langle X(t, x), \sqrt{C}\left(B(t + h) - B(t)\right)\rangle$$

$$+ 2\left\langle \int_t^{t+h} b(X(s, x))ds, \sqrt{C}\left(B(t + h) - B(t)\right)\right\rangle$$

Taking expectation we find that

$$\mathbb{E}\left[|X(t + h, x)|^2 - |X(t, x)|^2\right] = \mathbb{E}\left|\int_t^{t+h} b(X(s, x))ds\right|^2$$

$$+ \mathrm{Tr}\,[C] + 2\mathbb{E}\left\langle X(t, x), \int_t^{t+h} b(X(s, x))ds\right\rangle$$

$$+ 2\mathbb{E}\left\langle \int_t^{t+h} b(X(s, x))ds, \sqrt{C}\left(B(t + h) - B(t)\right)\right\rangle.$$

Now the conclusion follows, dividing both sides of the identity above by h and letting h tend to 0. $\square$

Remark 4.22 Usually the forward or backward Itô formulae are stated more generally for $\varphi(X(t, x, s))$ instead of for $\mathbb{E}[\varphi(X(t, x, s))]$. One can show that the stochastic process $u(t, s, x) = \varphi(X(t, x, s))$ is the solution to a stochastic partial differential equation.

4.5 A generalization

For several applications the assumption that b is Lipschitz continuous is too restrictive. Previous results can be generalized however under the following assumption.

Hypothesis 4.23 *(i) b is locally Lipschitz continuous.*
(ii) There exists a continuous non-negative function a such that

$$\langle b(x + y), x \rangle \le a(y)(1 + |x|^2), \quad \text{for all } x, y \in \mathbb{R}^n.$$

We can prove in fact the following result.

Proposition 4.24 *Assume that Hypothesis 4.23 holds. Let $x \in H$ and $f \in C([0, T]; H)$. Then the equation*

$$u(t) = x + \int_0^t b(u(s))ds + f(t), \quad t \in [0, T], \quad (4.54)$$

has a unique solution $u \in C([0, T]; H)$.

Proof. Setting $v = u - f$, equation (4.54) reduces to

$$v(t) = x + \int_0^t b(v(s) + f(s))ds,$$

which is equivalent to the Cauchy problem

$$\begin{cases} \dfrac{d}{dt} v(t) = b(v(t) + f(t)), \\[2mm] v(0) = x + f(0). \end{cases} \quad (4.55)$$

Since b is locally Lipschitz continuous, there exists a unique solution of (4.55) in a maximal interval $[0, \tau^*)$ included in $[0, T]$. To show global existence we have to find an a-priori estimate for v.

In fact, multiplying both sides of the first equation in (4.55) by $v(t)$, and taking into account Hypothesis (4.23)(ii), we find that

$$\frac{1}{2} \frac{d}{dt} |v(t)|^2 = \langle b(v(t) + f(t)), v(t) \rangle \le a(f(t))(1 + |v(t)|^2).$$

Set

$$\kappa = \sup_{t \in [0,T]} a(f(t)).$$

Then we have

$$\frac{1}{2} \frac{d}{dt} |v(t)|^2 \leq \kappa (1 + |v(t)|^2),$$

which yields

$$|v(t)|^2 \leq e^{2\kappa T} |x + f(0)|^2 + \int_0^T e^{2\kappa s} ds,$$

the required a-priori estimate. $\square$

Now several of the previous considerations can be generalized; we leave this task as an exercise to the reader.

Exercise 4.25 Let $n = 1$, $b(x) = -x^{2m+1}$ where $m \in \mathbb{N}$, $x \in \mathbb{R}$ and $f \in C([0,T];\mathbb{R})$. Prove that the equation

$$u(t) = x - \int_0^t u(s)^{2m+1} ds + f(t), \quad t \in [0,T], \tag{4.56}$$

has a unique solution $u \in C([0,T];H)$.

5

Invariant measures for Markov semigroups

We are given a Hilbert space H (inner product $\langle \cdot, \cdot \rangle$, norm $| \cdot |$). We shall use the following notations.

- $B(x, r)$ is the open ball in H with centre x and radius $r > 0$.
- $C_b(H)$ (resp. $B_b(H)$) is the Banach space of all uniformly continuous and bounded mappings (resp. Borel bounded mappings) $\varphi \colon H \to \mathbb{R}$ endowed with the norm

$$\|\varphi\|_0 = \sup_{x \in H} |\varphi(x)|.$$

- $L(C_b(H))$ (resp. $L(B_b(H))$) is the space of all linear bounded operators from $C_b(H)$ (resp. $B_b(H)$) into itself.
- $C_b^+(H)$ (resp. $B_b^+(H)$) represents the cone in $C_b(H)$ (resp. $C_b(H)$) consisting of all non-negative functions, and $\mathbf{1}$ the function on H identically equal to 1.
- $C_b(H)^*$ is the topological dual of $C_b(H)$.
- $\mathscr{P}(H)$ is the space of all probability measures on $(H, \mathscr{B}(H))$ where $\mathscr{B}(H)$ is the σ-algebra of all Borel subsets of H.
  There is a natural embedding of $\mathscr{P}(H)$ into $C_b(H)^*$. Namely, for any $\mu \in \mathscr{P}(H)$ we set

$$F_\mu(\varphi) = \int_H \varphi(x)\mu(dx), \quad \varphi \in C_b(H).$$

In the following we shall often identify μ with F_μ.

5.1 Markov semigroups

Definition 5.1 *A* Markov semigroup P_t *on* $B_b(H)$ *is a mapping*

$$[0, +\infty) \to L(B_b(H)), \quad t \mapsto P_t,$$

such that

(i) $P_0 = 1$, $P_{t+s} = P_t P_s$ *for all* $t, s \geq 0$.

(ii) *For any* $t \geq 0$ *and* $x \in H$ *there exists a probability measure* $\pi_t(x, \cdot) \in \mathscr{P}(H)$ *such that*

$$P_t \varphi(x) = \int_H \varphi(y) \pi_t(x, dy) \quad \text{for all } \varphi \in B_b(H). \tag{5.1}$$

(iii) *For any* $\varphi \in C_b(H)$ *(resp.* $B_b(H)$*) and* $x \in H$*, the mapping* $t \mapsto P_t \varphi(x)$ *is continuous (resp. Borel).*

Obviously, by (5.1) it follows that for $t = 0$,

$$\pi_0(x, \cdot) = \delta_x, \quad x \in H,$$

where δ_x is the Dirac measure at x.

We notice that in the literature one requires usually only (i) and (ii) in the definition of Markov semigroup P_t. In this case condition (iii) means that P_t is *stochastically continuous*, see e.g. [10].

Definition 5.2 *Let* P_t *be a Markov semigroup.*

(i) P_t *is* Feller *if* $P_t \varphi \in C_b(H)$ *for any* $\varphi \in C_b(H)$ *and any* $t \geq 0$.

(ii) P_t *is* strong Feller *if* $P_t \varphi \in C_b(H)$ *for any* $\varphi \in B_b(H)$ *and any* $t > 0$.

(iii) P_t *is* irreducible *if* $P_t \mathbf{1}_{B(x_0, r)}(x) > 0$ *for all* $x, x_0 \in H$, $r > 0$ *and any* $t \geq 0$.

Let us give some general properties of a Markov semigroup P_t. First, notice that by (5.1) we have $P_t \mathbf{1} = \mathbf{1}$ for all $t \geq 0$ and that P_t *preserves positivity*, that is $P_t \varphi \in B_b^+(H)$ for all $\varphi \in B_b^+(H)$.

Moreover, since, for any $\varphi \in C_b(H)$,

$$-\|\varphi\|_0 \leq \varphi(x) \leq \|\varphi\|_0, \quad x \in H,$$

we have

$$|P_t \varphi(x)| \leq \|\varphi\|_0, \quad x \in H.$$

Consequently $\|P_t\|_{L(B_b(H))} \leq 1$, for any $t \geq 0$. That is P_t is a semigroup of contractions on $B_b(H)$.

Let us give now some properties of the family of measures $\pi_t(x, \cdot)$ (called a *probability kernel*).

By (5.1) it follows that for any $E \in \mathscr{B}(H)$ we have

$$\pi_t(x, E) = P_t \mathbf{1}_E(x), \quad t \geq 0, \ x \in H. \tag{5.2}$$

Moreover, the following useful result holds.

Proposition 5.3 *For any* $t, s \geq 0$, $x \in H$ *and any* $E \in \mathcal{B}(H)$ *we have*

$$\pi_{t+s}(x, E) = \int_H \pi_s(y, E) \pi_t(x, dy). \tag{5.3}$$

Proof. We have in fact, taking into account the semigroup property of P_t, (5.2) and (5.1),

$$\pi_{t+s}(x, E) = P_{t+s} 1_E(x) = P_t \pi_s(\cdot, E)(x) = \int_H \pi_s(y, E) \pi_t(x, dy).$$

$\square$

Example 5.4 Let us consider the differential equation

$$\begin{cases} X'(t) = b(X(t)), \\ X(0) = x, \end{cases} \tag{5.4}$$

on $H = \mathbb{R}^n$ where $b \colon H \to H$ is Lipschitz continuous. As is well known, there exists a unique solution $X(t, x)$ of problem (5.4). Set

$$\pi_t(x, \cdot) = \delta_{X(t,x)}, \quad x \in \mathbb{R}^n.$$

Then it is easy to see that the transition semigroup

$$P_t \varphi(x) = \varphi(X(t, x)), \quad \varphi \in B_b(\mathbb{R}^n) \tag{5.5}$$

is a Markov semigroup.

Exercise 5.5 (i) Prove that semigroup P_t, defined by (5.5), is Feller. Is P_t strong Feller?
(ii) Prove that P_t is strongly continuous in $C_b(H)$ if and only if b is bounded.

Example 5.6 Let us consider the stochastic differential equation

$$\begin{cases} dX = b(X)dt + \sqrt{C} \, dB(t), \\ X(0) = x, \end{cases} \tag{5.6}$$

on $H = \mathbb{R}^n$ where B is a standard Brownian motion in a probability space $(\Omega, \mathscr{F}, \mathbb{P})$ with values in H, $b \colon H \to H$ is locally Lipschitz continuous, $C \in L(H)$ and Hypothesis 4.23 is fulfilled.

Then by Proposition 4.3 there exists a unique continuous stochastic process $X(\cdot, x)$, the solution of problem (5.6). Set

$$\pi_t(x, E) = (X(t, x)_\# \mathbb{P})(E), \quad x \in \mathbb{R}^n, \; E \in \mathscr{B}(\mathbb{R}^n).$$

Then the transition semigroup

$$P_t \varphi(x) = \mathbb{E}\left[\varphi(X(t, x))\right] = \int_{\mathbb{R}} \varphi(y) \pi_t(x, dy), \quad \varphi \in B_b(H), \qquad (5.7)$$

is a Markov semigroup as easily checked.

Exercise 5.7 Prove that the semigroup P_t, defined by (5.7), is Feller.

5.2 Invariant measures

In this section P_t represents a Markov semigroup on H. A probability measure $\mu \in \mathscr{P}(H)$ is said to be *invariant* for P_t if

$$\int_H P_t \varphi d\mu = \int_H \varphi d\mu \quad \text{for all } \varphi \in B_b(H) \text{ and } t \geq 0. \qquad (5.8)$$

If P_t is Feller this condition is clearly equivalent (identifying μ with F_μ) to

$$P_t^* \mu = \mu \quad \text{for all } t \geq 0, \qquad (5.9)$$

where P_t^* is the transpose operator of P_t, defined as

$$\langle \varphi, P_t^* F \rangle = \langle P_t \varphi, F \rangle,$$

for all $\varphi \in C_b(H)$, $F \in C_b(H)^*$. [1]

If $\mu \in \mathscr{P}(H)$ is invariant for P_t we have

$$\mu(A) = P_t^* \mu(A) = \int_H P_t 1_A(x) \mu(dx), \quad A \in \mathscr{B}(H),$$

from which, recalling (5.8),

$$\mu(A) = \int_H \pi_t(x, A) \mu(dx), \quad A \in \mathscr{B}(H). \qquad (5.10)$$

A first basic result is the following.

---

[1] $\langle \cdot, \cdot \rangle$ represent the duality between $C_b(H)$ and $C_b(H)^*$.

Theorem 5.8 *Assume that μ is an invariant measure for P_t. Then for all $t \geq 0$, $p \geq 1$, P_t is uniquely extendible to a linear bounded operator on $L^p(H, \mu)$ that we still denote by P_t. Moreover*

$$\|P_t\|_{L(L^p(H,\mu))} \leq 1, \quad t \geq 0. \tag{5.11}$$

Finally, P_t is a strongly continuous semigroup in $L^p(H, \mu)$.

Proof. Let $\varphi \in C_b(H)$. By the Hölder inequality we have

$$|P_t\varphi(x)|^p \leq \int_H |\varphi(y)|^p \pi_t(x, dy) = P_t(|\varphi|^p)(x).$$

Integrating both sides of the above inequality with respect to μ over H yields

$$\int_H |P_t\varphi(x)|^p \mu(dx) \leq \int_H P_t(|\varphi|^p)(x)\mu(dx) = \int_H |\varphi(x)|^p \mu(dx)$$

in view of the invariance of μ. Since $C_b(H)$ is dense in $L^p(H, \mu)$, P_t is uniquely extendible to $L^p(H, \mu)$ and (5.11) follows.

Let us show finally that P_t is strongly continuous in $L^p(H, \mu)$. First let $\varphi \in C_b(H)$. Then, by property (iii) in Definition 5.1 of P_t we have that the function $t \to P_t\varphi(x)$ is continuous for any $x \in H$. Consequently, by the dominated convergence theorem

$$\lim_{t \to 0} P_t\varphi = \varphi \quad \text{in } L^p(H, \mu).$$

The same assertion follows easily when $\varphi \in L^p(H, \mu)$ by the density of $C_b(H)$ in $L^p(H, \mu)$. $\square$

Let μ be an invariant measure for P_t. We are going to study the asymptotic behaviour of $P_t\varphi$, for $\varphi \in L^2(H, \mu)$. This is obvious when $P_t\varphi = \varphi$ for all $t > 0$. In this case we say that φ is *stationary*. In general, given $\varphi \in L^2(H, \mu)$, one can ask whether there exists the limit

$$\lim_{t \to +\infty} P_t\varphi(x), \tag{5.12}$$

or, if not, if there exists the limit of the means

$$\lim_{T \to +\infty} \frac{1}{T} \int_0^T P_s\varphi(x)ds. \tag{5.13}$$

We shall prove indeed that this limit always exists in $L^2(H, \mu)$ (*Von Neumann theorem*).

If in addition it happens that

$$\lim_{T \to +\infty} \frac{1}{T} \int_0^T P_t\varphi(x)dt = \int_H \varphi d\mu \quad \text{in } L^2(H, \mu), \tag{5.14}$$

for all $\varphi \in L^2(H, \mu)$, P_t is said to be *ergodic*. In this case the identity (5.14) is interpreted in physics by saying that the "temporal" average of $P_t\varphi$ coincides with the "spatial" average of φ.

It can also happen in particular that

$$\lim_{t \to +\infty} P_t\varphi(x) = \int_H \varphi d\mu \quad \text{in } L^2(H, \mu). \tag{5.15}$$

In this case P_t is said to be *strongly mixing*.

Existence and uniqueness of invariant measures will be proved in Chapter 7. We conclude this introduction by giving two examples of invariant measures.

Exercise 5.9 Consider the ordinary differential equation,

$$Z'(t) = Z(t) - Z^3(t), \quad Z(0) = x,$$

and the corresponding transition semigroup

$$P_t\varphi(x) = \varphi(Z(t, x)), \quad \varphi \in C_b(H).$$

Prove that P_t is a Markov semigroup and that $\pi_t(x, E) = \delta_{Z(t,x)}(E)$, $E \in \mathscr{B}(\mathbb{R})$, $t \geq 0$, $x \in \mathbb{R}$.

Show moreover that measures δ_0, δ_1 and δ_{-1} are invariant, ergodic and strongly mixing.

Exercise 5.10 Consider the stochastic differential equation in $\mathbb{R}$,

$$dX(t) = -X(t)dt + dB(t), \quad X(0) = x,$$

whose solution $X(t, x)$ is given by the Ornstein–Uhlenbeck process (see Proposition 4.10),

$$X(t, x) = e^{-t}x + \int_0^t e^{-(t-s)}dB(s), \quad t \geq 0, \ x \in \mathbb{R}.$$

Prove that

$$\pi_t(x, \cdot) = N_{e^{-t}x, \frac{1}{2}(1-e^{-2t})}, \quad x \in \mathbb{R}, \ t > 0.$$

Show moreover that the measure $\mu = N_{\frac{1}{2}}$ is invariant, ergodic and strongly mixing.

Hint. Check that (5.8) holds for $\varphi(x) = e^{ihx}$, where $h \in \mathbb{R}$.

In order to study the behaviour of $\lim_{T \to +\infty} \frac{1}{T} \int_0^T P_t \varphi \, dt$, we need some general result about the averages of the powers of a linear operator, proved in the next section.

5.3 Ergodic averages

We are given a linear bounded operator T on a Hilbert space E (norm $\| \cdot \|$, inner product $\langle \cdot, \cdot \rangle$).[2] We set

$$M_n = \frac{1}{n} \sum_{k=0}^{n-1} T^k, \quad n \in \mathbb{N}.$$

Theorem 5.11 *Assume that $\sup_{n \in \mathbb{N}} \|T^n\| < +\infty$. Then there exists the limit*

$$\lim_{n \to \infty} M_n x := M_\infty x \quad \text{for all } x \in E. \tag{5.16}$$

Moreover $M_\infty \in L(H)$, $M_\infty^2 = M_\infty$ and $M_\infty(E) = \text{Ker} \, (1 - T)$.

Proof. First notice that the limit of $(M_n x)$ certainly exists when either $x \in \text{Ker} \, (1 - T)$, or $x \in (1 - T)(E)$. In fact in the first case we have obviously

$$\lim_{n \to \infty} M_n x = x \quad \text{for all } x \in \text{Ker} \, (1 - T),$$

and in the latter we have

$$\lim_{n \to \infty} M_n x = 0 \quad \text{for all } x \in (1 - T)(E),$$

because

$$(1 - T)M_n = M_n(1 - T) = \frac{1}{n}(1 - T^n), \quad n \in \mathbb{N}. \tag{5.17}$$

Consequently we also have

$$\lim_{n \to \infty} M_n x = 0 \quad \text{for all } x \in \overline{(1 - T)(E)}, \tag{5.18}$$

where $\overline{(1 - T)(E)}$ is the closure of $(1 - T)(E)$.

Now let $x \in E$ be fixed. Since $\|M_n x\|_{n \in \mathbb{N}}$ is bounded by assumption, there exists a sub-sequence (n_k) of $\mathbb{N}$, and an element $y \in H$ such that $M_{n_k} x \to y$ weakly as $k \to \infty$. By (5.17) it follows also that $T M_{n_k} x \to Ty = y$, so that $y \in \text{Ker} \, (1 - T)$.

_____

[2] Later we shall take $E = L^2(H, \mu)$.

Now we prove that $M_n x \to y$. First note that, since $y \in \mathrm{Ker}\,(1-T)$, we have $M_n y = y$, and so

$$M_n x = M_n y + M_n(x - y) = y + M_n(x - y). \tag{5.19}$$

We claim that $x - y \in \overline{(1 - T)(E)}$, which will prove (5.17) by (5.16). We have in fact

$$x - y = \lim_{k \to \infty} (x - M_{n_k} x),$$

and $x - M_{n_k} x \in (1 - T)(E)$ because

$$x - M_{n_k} x = \frac{1}{n_k} \sum_{h=0}^{n_k - 1} (1 - T^h) x$$

$$= \frac{1}{n_k} \sum_{h=0}^{n_k - 1} (1 + T + \ldots + T^{h-1})(1 - T) x.$$

Therefore (5.16) holds.

Finally, since $(1 - T)M_n \to 0$, we have $M^\infty = TM^\infty$, so that $T^k M^\infty = M^\infty$, $k \in \mathbb{N}$, and $M^\infty = M_n M^\infty$, which yields as $n \to \infty$, $M^\infty = (M^\infty)^2$, as required. $\square$

5.4 The Von Neumann theorem

In this section we assume that there is an invariant measure μ for the Markov semigroup P_t. This will allow us to extend the semigroup P_t to $L^2(H, \mu)$, as proved in Theorem 5.8.

We denote by Σ the set

$$\Sigma = \{ f \in L^2(H, \mu) : P_t f = f, \ \mu\text{-a.e. for all } t \geq 0 \} \tag{5.20}$$

of all *stationary* points of P_t. Clearly Σ is a closed subspace of $L^2(H, \mu)$ and $\mathbf{1} \in \Sigma$.

Let us consider the average

$$M(T)\varphi = \frac{1}{T} \int_0^T P_t \varphi \, dt, \quad \varphi \in L^2(H, \mu), \ T > 0.$$

Theorem 5.12 *There exists the limit*

$$\lim_{T \to \infty} M(T)\varphi =: M_\infty \varphi \quad \text{in } L^2(H, \mu). \tag{5.21}$$

Moreover M_∞ is a projection operator on Σ, and

$$\int_H M_\infty \varphi d\mu = \int_H \varphi d\mu. \tag{5.22}$$

Proof. For all $T > 0$ write

$$T = n_T + r_T, \quad n_T \in \mathbb{N} \cup \{0\}, \ r_T \in [0, 1).$$

For $\varphi \in L^2(H, \mu)$ we have

$$M(T)\varphi = \frac{1}{T} \sum_{k=0}^{n_T-1} \int_k^{k+1} P_s \varphi ds + \frac{1}{T} \int_{n_T}^T P_s \varphi ds$$

$$= \frac{1}{T} \sum_{k=0}^{n_T-1} \int_0^1 P_{s+k} \varphi ds + \frac{1}{T} \int_0^{r_T} P_{s+n(T)} \varphi ds$$

$$= \frac{n_T}{T} \frac{1}{n_T} \sum_{k=0}^{n_T-1} (P_1)^k M(1)\varphi + \frac{r_T}{T}(P_1)^{n_T} M(r_T)\varphi. \tag{5.23}$$

Since

$$\lim_{T\to\infty} \frac{n_T}{T} = 1, \quad \lim_{T\to\infty} \frac{r_T}{T} = 0,$$

letting $n \to \infty$ in (5.23) and invoking Theorem 5.11, we get (5.21).

We prove now that for all $t \geq 0$

$$M_\infty P_t = P_t M_\infty = M_\infty. \tag{5.24}$$

In fact, given $t \geq 0$ we have

$$M_\infty P_t \varphi = \lim_{T\to\infty} \frac{1}{T} \int_0^T P_{t+s} \varphi ds = \lim_{T\to\infty} \frac{1}{T} \int_t^{t+T} P_s \varphi ds$$

$$= \lim_{T\to\infty} \frac{1}{T} \left\{ \int_0^T P_s \varphi ds - \int_0^t P_s \varphi ds + \int_T^{T+t} P_s \varphi ds \right\}$$

$$= M_\infty \varphi$$

and this yields (5.24).

By (5.24) it follows that $M_\infty f \in \Sigma$ for all $f \in L^2(H, \mu)$, and moreover that

$$M_\infty M(T) = M(T) P_\infty = M_\infty,$$

which yields, letting $T \to \infty$, $M_\infty^2 = M_\infty$. Finally, (5.22) follows, by integrating (5.21) with respect to μ. $\square$

5.5 Ergodicity

Let μ be an invariant measure for P_t. We say that μ is *ergodic* if

$$\lim_{T \to \infty} \frac{1}{T} \int_0^T P_t \varphi \, dt = \overline{\varphi} \quad \text{for all } \varphi \in L^2(H, \mu), \tag{5.25}$$

where

$$\overline{\varphi} = \int_H \varphi(x) \mu(dx).$$

Proposition 5.13 *Let μ be an invariant measure for P_t. Then μ is ergodic if and only if the dimension of the linear space Σ of all stationary elements of $L^2(H, \mu)$ defined by (5.20) is 1.*

Proof. If μ is ergodic it follows from (5.25) that any element in Σ is constant, so that dimension of Σ is 1. Conversely assume that dimension of Σ is 1. Then there is a linear bounded functional F on $L^2(H, \mu)$ such that

$$M_\infty \varphi = F(\varphi) \mathbf{1}.$$

By the Riesz representation theorem there exists an element $\varphi_0 \in L^2(H, \mu)$ such that $F(\varphi) = \langle \varphi, \varphi_0 \rangle$. Integrating this equality on H with respect to μ and taking into account the invariance of M_∞ (see (5.22)), yields

$$\int_H M_\infty \varphi \, d\mu = \int_H \varphi \, d\mu = \langle \varphi, \mathbf{1} \rangle = \langle \varphi, \varphi_0 \rangle, \quad \varphi \in L^2(H, \mu).$$

Therefore $\varphi_0 = \mathbf{1}$. $\square$

Let μ be an invariant measure for P_t. A Borel set $\Gamma \in \mathcal{B}(H)$ is said to be *invariant* for P_t if its characteristic function $\mathbf{1}_\Gamma$ belongs to Σ. If $\mu(\Gamma)$ is equal to either 0 or 1, we say that Γ is *trivial*, otherwise it is *nontrivial*.

We now want to show that μ is ergodic if and only if all invariant sets are trivial. For this it is important to notice that Σ is a lattice, as proved in the next proposition.

Proposition 5.14 *Assume that φ and ψ belong to Σ. Then the following statements hold.*

(i) $|\varphi| \in \Sigma$.
(ii) $\varphi^+, \varphi^- \in \Sigma$. [3]

---

[3] $\varphi^+ = \max\{\varphi, 0\}$, $\varphi^- = \max\{-\varphi, 0\}$.

(iii) $\varphi \vee \psi$, $\varphi \wedge \psi \in \Sigma$. [4]

(iv) For any $a \in \mathbb{R}$ we have $\mathbf{1}_{\{x \in H: \ \varphi(x) > a\}} \in \Sigma$.

Proof. Let us prove (i). Let $t > 0$ and assume that $\varphi \in \Sigma$, so that $\varphi(x) = P_t \varphi(x)$. Then we have

$$|\varphi(x)| = |P_t \varphi(x)| \leq P_t(|\varphi|)(x), \quad x \in H. \tag{5.26}$$

We claim that

$$|\varphi(x)| = P_t(|\varphi|)(x), \quad \mu\text{-a.s.}$$

Assume by contradiction that there is a Borel subset $I \subset H$ such that $\mu(I) > 0$ and

$$|\varphi(x)| < P_t(|\varphi|)(x), \quad x \in I.$$

Then we have

$$\int_H |\varphi(x)| \mu(dx) < \int_H P_t(|\varphi|)(x) \mu(dx).$$

Since, by the invariance of μ,

$$\int_H P_t(|\varphi|)(x) \mu(dx) = \int_H |\varphi|(x) \mu(dx),$$

we find a contradiction.

Statements (ii) and (iii) follow from the obvious identities

$$\varphi^+ = \frac{1}{2}(\varphi + |\varphi|), \quad \varphi^- = \frac{1}{2}(\varphi - |\varphi|),$$

$$\varphi \vee \psi = (\varphi - \psi)^+ + \psi, \quad \varphi \wedge \psi = -(\varphi - \psi)^+ + \varphi.$$

Finally let us prove (iv). It is enough to show that the set $\{\varphi > 0\}$ is invariant, or, equivalently, that $\mathbf{1}_{\{\varphi > 0\}}$ belongs to Σ. We have in fact, as it is easily checked,

$$\mathbf{1}_{\{\varphi > 0\}} = \lim_{n \to \infty} \varphi_n(x), \quad x \in H,$$

where $\varphi_n = (n\varphi^+) \wedge \mathbf{1}$, $n \in \mathbb{N}$, belongs to Σ by (ii) and (iii). Therefore $\{\varphi > 0\}$ is invariant. $\square$

We are now ready to prove the following result.

Theorem 5.15 *Let μ be an invariant measure for P_t. Then μ is ergodic if and only if any invariant set is trivial.*

---

[4] $\varphi \vee \psi = \max\{\varphi, \psi\}$, $\varphi \wedge \psi = \min\{\varphi, \psi\}$.

Proof. Let Γ be invariant for μ. Then if μ is ergodic $\mathbf{1}_\Gamma$ must be constant (otherwise dim $\Sigma \geq 2$) and so Γ is trivial. Assume conversely that the only invariant sets for μ are trivial and, by contradiction, that μ is not ergodic. Then there exists a non-constant function $\varphi_0 \in \Sigma$. Therefore by Proposition 5.14 for some $\lambda \in \mathbb{R}$ the invariant set $\{\varphi_0 > \lambda\}$ is not trivial. $\square$

5.6 Structure of the set of all invariant measures

We still assume that P_t is a Markov semigroup on H. We denote by Λ the set of all its invariant measures and we assume that Λ is non-empty. Clearly Λ is a convex subset of $C_b(H)^*$.

Theorem 5.16 *Assume that there is a unique invariant measure μ for P_t. Then μ is ergodic.*

Proof. Assume by contradiction that μ is not ergodic. Then there is a nontrivial invariant set Γ. Let us prove that the measure μ_Γ defined as

$$\mu_\Gamma(A) = \frac{1}{\mu(\Gamma)} \, \mu(A \cap \Gamma), \quad A \in \mathscr{B}(H),$$

belongs to Λ. This will give rise to a contradiction.
Recalling (5.10), we have to show that

$$\mu_\Gamma(A) = \int_H \pi_t(x, A) \mu_\Gamma(dx), \quad A \in \mathscr{B}(H),$$

or, equivalently, that

$$\mu(A \cap \Gamma) = \int_\Gamma \pi_t(x, A) \mu(dx), \quad A \in \mathscr{B}(H). \tag{5.27}$$

In fact, since Γ is invariant, we have

$$P_t \mathbf{1}_\Gamma = \mathbf{1}_\Gamma, \quad P_t \mathbf{1}_{\Gamma^c} = \mathbf{1}_{\Gamma^c}, \quad t \geq 0,$$

and so
$$\pi_t(x, \Gamma) = \mathbf{1}_\Gamma(x), \quad \pi_t(x, \Gamma^c) = \mathbf{1}_{\Gamma^c}(x), \quad t \geq 0.$$

Consequently,

$$\pi_t(x, A \cap \Gamma^c) = 0, \quad \mu\text{-a.e. in } \Gamma \text{ and } \pi_t(x, A \cap \Gamma) = 0, \quad \mu\text{-a.e. in } \Gamma^c,$$

and so

$$\int_\Gamma \pi_t(x, A)\mu(dx) = \int_\Gamma \pi_t(x, A \cap \Gamma)\mu(dx) + \int_\Gamma \pi_t(x, A \cap \Gamma^c)\mu(dx)$$

$$= \int_\Gamma \pi_t(x, A \cap \Gamma)\mu(dx) = \int_H \pi_t(x, A \cap \Gamma)\mu(dx) = \mu(A \cap \Gamma),$$

and (5.10) holds. $\square$

We want now to show that the set of all extremal points of Λ is precisely the set of all ergodic measures of P_t. For this we need a lemma.

Lemma 5.17 *Let* $\mu, \nu \in \Lambda$ *with* μ *ergodic and* ν *absolutely continuous with respect to* μ. *Then* $\mu = \nu$.

Proof. Let $\Gamma \in \mathscr{B}(H)$. By the Von Neumann theorem there exists $T_n \uparrow \infty$ such that

$$\lim_{n \to \infty} \frac{1}{T_n} \int_0^{T_n} P_t 1_\Gamma dt = \mu(\Gamma), \quad \mu\text{-a.e.} \tag{5.28}$$

Since $\nu \ll \mu$, identity (5.28) holds also ν-a.e. Now integrating (5.28) with respect to ν yields

$$\lim_{n \to \infty} \frac{1}{T_n} \int_0^{T_n} \left(\int_H P_t 1_\Gamma d\nu \right) dt = \nu(\Gamma), \quad \mu\text{-a.e.}$$

Consequently $\nu(\Gamma) = \mu(\Gamma)$ as required. $\square$

We can now prove the announced property of Λ.

Theorem 5.18 *The set of all invariant ergodic measures of* P_t *coincides with the set of all extremal points of* Λ.

Proof. We first prove that if μ is ergodic then it is an extremal point of Λ. Assume by contradiction that μ is ergodic and it is not an extremal point of Λ. Then there exist $\mu_1, \mu_2 \in \Lambda$ with $\mu_1 \neq \mu_2$, and $\alpha \in (0, 1)$ such that

$$\mu = \alpha \mu_1 + (1 - \alpha)\mu_2.$$

Then clearly $\mu_1 \ll \mu$ and $\mu_2 \ll \mu$. By Lemma 5.17 we get a contradiction.

We finally prove that if μ is an extremal point of Λ, then it is ergodic. Assume by contradiction that μ is not ergodic. Then there exists a nontrivial invariant set Γ. Consequently, arguing as in the proof of Theorem 5.16, we have $\mu_\Gamma, \mu_{\Gamma^c} \in \Lambda$. Since

$$\mu = \mu(\Gamma)\mu_\Gamma + (1 - \mu(\Gamma))\mu_{\Gamma^c},$$

we find that μ is not extremal, a contradiction. $\square$

Theorem 5.19 *Assume that μ and ν are ergodic invariant measures with $\mu \neq \nu$. Then μ and ν are singular.*

Proof. Let $\Gamma \in \mathscr{B}(H)$ be such that $\mu(\Gamma) \neq \nu(\Gamma)$. From the Von Neumann theorem it follows that there exists $T_n \uparrow +\infty$ and two Borel sets M and N such that $\mu(M) = 1, \nu(N) = 1$, and

$$\lim_{n \to \infty} \frac{1}{T_n} \int_0^{T_n} (P_t \mathbf{1}_\Gamma)(x)dt = \mu(\Gamma), \ \forall \, x \in M,$$

$$\lim_{n \to \infty} \frac{1}{T_n} \int_0^{T_n} (P_t \mathbf{1}_\Gamma)(x)dt = \nu(\Gamma), \ \forall \, x \in N.$$

Since $\mu(\Gamma) \neq \nu(\Gamma)$ this implies that $M \cap N = \varnothing$, and so μ and ν are singular. $\square$

6

Weak convergence of measures

We are given a separable Hilbert space H (norm $|\cdot|$, inner product $\langle \cdot, \cdot \rangle$). We shall use notations from the previous chapters. In particular, $\mathscr{B}(H)$ is the σ-algebra of all Borel subsets of H and $\mathscr{P}(H)$ is the set of all probability measures on $(H, \mathscr{B}(H))$. If $B \in \mathscr{B}(H)$ we denote as before by B^c its complement.

Moreover, $C_b(H)$ is the Banach space of all uniformly continuous and bounded mappings $\varphi \colon H \to \mathbb{R}$ endowed with the sup norm

$$\|\varphi\|_0 = \sup_{x \in H} |\varphi(x)|.$$

$C_b(H)^*$ is the topological dual of $C_b(H)$, endowed with the norm

$$\|F\| = \sup \{|F(x)| : |x| < 1\}, \quad F \in C_b(H)^*.$$

We follow here [19].

6.1 Some additional properties of measures

Let us first prove that any Borel probability measure on H is *regular*.

Proposition 6.1 *Let $\mu \in \mathscr{P}(H)$. Then for any $B \in \mathscr{B}(H)$ we have*

$$\mu(B) = \sup\{\mu(C) : C \subset B, \ \text{closed}\} = \inf\{\mu(A) : A \supset B, \ \text{open}\}. \tag{6.1}$$

Proof. Let us set

$$\mathscr{K} = \{B \in \mathscr{B}(H) : (6.1) \ \text{holds}\}.$$

Obviously $\mathscr{K}$ contains H and $\varnothing$. Thus it is enough to show that $\mathscr{K}$ is a σ-algebra of parts of H including the open sets of H. Clearly if

$B \in \mathcal{K}$ then its complement B^c belongs to $\mathcal{K}$. Let us prove now that $(B_n) \subset \mathcal{K} \Rightarrow \bigcup_{n=1}^{\infty} B_n \in \mathcal{K}$. Fix $\varepsilon > 0$. We are going to show that there is a closed set C and an open set A such that

$$C \subset \bigcup_{n=1}^{\infty} B_n \subset A, \quad \mu(A \backslash C) \leq \varepsilon. \tag{6.2}$$

Let $n \in \mathbb{N}$. Since $B_n \in \mathcal{K}$ there is an open set A_n and a closed set C_n such that

$$C_n \subset B_n \subset A_n : \mu(A_n \backslash C_n) \leq \frac{\varepsilon}{2^{n+1}}.$$

Setting

$$A = \bigcup_{n=1}^{\infty} A_n; \ S = \bigcup_{n=1}^{\infty} C_n$$

we have $S \subset \bigcup_{n=1}^{\infty} B_n \subset A$ and $\mu(A \backslash S) \leq \frac{\varepsilon}{2}$. However, A is open but S is not necessarily closed. So we approximate S by closed sets setting $S_n = \bigcup_{k=1}^{n} C_k$. S_n is obviously closed, $S_n \uparrow S$ and consequently $\mu(S_n) \uparrow \mu(S)$. Therefore there exists $n_\varepsilon \in \mathbb{N}$ such that $\mu(S \backslash S_{n_\varepsilon}) < \frac{\varepsilon}{2}$. Now setting $C = S_{n_\varepsilon}$ we have $C \subset \bigcup_{n=1}^{\infty} B_n \subset A$ and $\mu(A \backslash C) < \varepsilon$. Therefore $\bigcup_{n=1}^{\infty} B_n \in \mathcal{K}$. We have proved that $\mathcal{K}$ is a σ-algebra. It remains to show that $\mathcal{K}$ contains the open subsets of H. In fact, let A be open and set

$$C_n = \left\{ x \in H : d(x, A^c) \geq \frac{1}{n} \right\},$$

where $d(x, A^c)$ represents the distance from x to A^c. Then C_n is a closed subsets of A, and moreover $C_n \uparrow A$ which implies $\mu(A/C_n) \downarrow 0$. Thus the conclusion follows. $\square$

Now we show that any probability measure $\mu \in \mathscr{P}(H)$ is concentrated on the union of a countable family of compact subsets of H. We have in fact the following result.

Theorem 6.2 *Let $\mu \in \mathscr{P}(H)$. Then there exists an increasing sequence of compact subsets (K_n) of H such that*

$$\lim_{n \to \infty} \mu(K_n) = 1.$$

Proof. Let $(a_h)_{h \in \mathbb{N}}$ be a dense subset of H. Then for any $k \in \mathbb{N}$ we have

$$H = \bigcup_{h=1}^{\infty} \overline{B(a_h, 1/k)},$$

where $\overline{B(a_h,1/k)} = \{y \in H : |y - a_h| \leq 1/k\}$. Consequently, for all $j, k \in \mathbb{N}$ there exists $n_{k,j} \in \mathbb{N}$ such that

$$\mu\left(\bigcup_{h=1}^{n_{k,j}} \overline{B(a_h, 1/k)}\right) \geq 1 - \frac{1}{j2^k},$$

or, equivalently,

$$\mu\left(\bigcap_{h=1}^{n_{k,j}} \overline{B(a_h, 1/k)}^{\,c}\right) \leq \frac{1}{j2^k}. \tag{6.3}$$

Set now,

$$K_j = \bigcap_{k=1}^{\infty} \bigcup_{h=1}^{n_{k,j}} \overline{B(a_h, 1/k)}.$$

Clearly, K_j can be covered by a finite number of balls of arbitrary radius so that is totally bounded and consequently relatively compact.[1] It remains to check that $\mu(K_j) \geq 1 - \frac{1}{j}$, which will conclude the proof. We have in fact taking into account (6.3),

$$\mu(K_j^c) = \mu\left(\bigcup_{k=1}^{\infty} \bigcap_{h=1}^{n_{k,j}} \overline{B(a_h, 1/k)}^{\,c}\right)$$

$$\leq \sum_{k=1}^{\infty} \mu\left(\bigcap_{h=1}^{n_{k,j}} \overline{B(a_h, 1/k)}^{\,c}\right) \leq \frac{1}{j}\sum_{k=1}^{\infty}\frac{1}{2^k} = \frac{1}{j}.$$

$\square$

6.2 Positive functionals

A linear functional $F \colon C_b(H) \to \mathbb{R}$ is said to be *positive*, if

$$\varphi \geq 0 \implies F(\varphi) \geq 0.$$

A linear positive functional F is continuous and we have $\|F\| = F(1)$. In fact from the inequalities

$$-\|\varphi\|_0 \leq \varphi(x) \leq \|\varphi\|_0, \quad x \in H, \ \varphi \in C_b(H),$$

it follows that

$$-\|\varphi\|_0 F(1) \leq F(\varphi) \leq \|\varphi\|_0 F(1), \quad \varphi \in C_b(H).$$

---

[1] See [14, page 22].

To any $\mu \in \mathscr{P}(H)$ we associate as before the positive functional

$$F_\mu(\varphi) = \int_H \varphi(x)\mu(dx), \quad \varphi \in C_b(H)$$

and we identify F_μ with μ.

By Proposition 1.5, the application $\mu \mapsto F_\mu$ is one-to-one. Obviously, it is not onto. We are going to characterize those positive functionals F such that $F = F_\mu$ for some measure μ. We start with a straightforward result.

Proposition 6.3 *Let F be a linear positive functional such that $F(\mathbf{1}) = 1$. If there exists $\mu \in \mathscr{P}(H)$ such that either*

$$F(\varphi) \le \int_H \varphi(x)\mu(dx), \quad \text{for all } \varphi \in C_b(H) \text{ with } 0 \le \varphi \le 1, \quad (6.4)$$

or

$$F(\varphi) \ge \int_H \varphi(x)\mu(dx), \quad \text{for all } \varphi \in C_b(H) \text{ with } 0 \le \varphi \le 1, \quad (6.5)$$

then $F = \mu$.

Proof. Assume for instance that (6.4) holds. Then we have

$$F(1 - \varphi) \le \int_H (1 - \varphi(x))\mu(dx) \quad \text{for all } \varphi \in C_b(H) \text{ with } 0 \le \varphi \le 1,$$

which, together with (6.4), yields $F = F_\mu$. $\square$

We want now to show that the restriction F_K (defined by (6.6)) of a positive functional F to a compact set K, is a measure. For any non-negative $\varphi \in C_b(H)$ we set

$$F_K(\varphi) = \inf \{F(\psi) : \psi \in C_b(H), \ \psi \ge 0, \ \psi = \varphi \text{ on } K\} \quad (6.6)$$

and

$$F_K(\varphi) = F_K(\varphi^+) - F_K(\varphi^-),$$

where

$$\varphi^+ = \max\{\varphi, 0\}, \quad \varphi^- = \max\{-\varphi, 0\}.$$

Obviously

$$F_K(\varphi) \le F(\varphi) \quad \text{for all non-negative } \varphi \in C_b(H). \quad (6.7)$$

Lemma 6.4 *F_K is linear.*

Proof. It is enough to show that if $\varphi_1, \varphi_2 \in C_b(H)$ are non-negative, we have

$$F_K(\varphi_1 + \varphi_2) = F_K(\varphi_1) + F_K(\varphi_2). \tag{6.8}$$

For any $\varepsilon > 0$ there exist $\psi_{1,\varepsilon}, \psi_{2,\varepsilon} \geq 0$ in $C_b(H)$ such that

$$\psi_{1,\varepsilon}(x) = \varphi_1(x), \quad \psi_{2,\varepsilon}(x) = \varphi_2(x), \quad x \in K$$

and

$$F_K(\varphi_i) + \varepsilon \geq F(\psi_{i,\varepsilon}), \quad i = 1, 2.$$

Therefore

$$F_K(\varphi_1 + \varphi_2) \leq F(\psi_{1,\varepsilon} + \psi_{1,\varepsilon}) = F(\psi_{1,\varepsilon}) + F(\psi_{2,\varepsilon}) \leq F_K(\varphi_1) + F_K(\varphi_2) + 2\varepsilon.$$

Since ε is arbitrary, we obtain

$$F_K(\varphi_1 + \varphi_2) \leq F_K(\varphi_1) + F_K(\varphi_2). \tag{6.9}$$

Conversely, given $\varepsilon > 0$ there is $\psi_\varepsilon \geq 0$ such that $\psi_\varepsilon = \varphi_1 + \varphi_2$ on K, and

$$F(\psi_\varepsilon) \leq F_K(\varphi_1 + \varphi_2) + \varepsilon.$$

We may choose ψ_ε such that

$$\psi_\varepsilon(x) \leq \varphi_1(x) + \varphi_2(x), \quad x \in H.$$

Set now

$$\psi_{i,\varepsilon}(x) = \begin{cases} \dfrac{\varphi_i(x)}{\varphi_1(x) + \varphi_2(x)} \, \psi_\varepsilon(x) & \text{if } \varphi_1(x) + \varphi_2(x) > 0, \\ 0 & \text{if } \varphi_1(x) + \varphi_2(x) = 0. \end{cases} \quad i = 1, 2.$$

Functions $\psi_{i,\varepsilon}$ are continuous since

$$\psi_{i,\varepsilon}(x) \leq \psi_\varepsilon(x) \leq \varphi_1(x) + \varphi_2(x), \quad x \in H.$$

Finally, we have

$$F_K(\varphi_1) + F_K(\varphi_2) \leq F(\psi_{1,\varepsilon}) + F(\psi_{2,\varepsilon}) = F(\psi_\varepsilon) \leq F_K(\varphi_1 + \varphi_2) + \varepsilon. \tag{6.10}$$

The conclusion follows from (6.9) and (6.10). $\square$

Proposition 6.5 *Let F be a positive functional on $C_b(H)$, and let K be a compact subset of H. Then there exists a measure $\mu_K \in \mathscr{P}(H)$ such that*

$$F_K(\varphi) = \int_H \varphi(x) \mu_k(dx), \quad \varphi \in C_b(H), \tag{6.11}$$

where F_K is defined by (6.6).

Proof. We first remark that, in virtue of the Tietze theorem (see [14, page 15]) any function of $C(K)$ is the restriction to K of a function of $C_b(H)$. Then we can define a positive functional $\tilde{F}_K$ on $C(K)$ setting

$$\tilde{F}_K(\varphi) = F_K(\tilde{\varphi}), \quad \varphi \in C(K),$$

where $\tilde{\varphi}$ is any non-negative function belonging to $C_b(H)$ which extends φ.

In virtue of the Riesz representation theorem (see [14, page 265]) there exists a unique Borel measure $\tilde{\mu}_K$ on K such that

$$\tilde{F}_K(\varphi) = \int_K \varphi(y)\tilde{\mu}_K(dy), \quad \varphi \in C(K).$$

Setting now

$$\mu_K(B) = \tilde{\mu}_K(B \cap K), \quad B \in \mathscr{B}(H),$$

we have that μ_K is a Borel measure (not a probability measure however in general) on H. Moreover

$$F_K(\varphi) = \int_K \varphi(x)\tilde{\mu}_K(dx) = \int_H \varphi(x)\mu_K(dx), \quad \varphi \in C_b(H),$$

and (6.11) follows. $\square$

Before proving the required characterization of those positive functionals which are measures, we need a definition. We say that a positive linear functional on H such that $F(\mathbf{1}) = 1$ is *tight* if there exists an increasing sequence of compact subsets (K_n) of H such that

$$\lim_{n \to \infty} F_{K_n}(\mathbf{1}) = 1.$$

Theorem 6.6 *Let F be a positive linear functional on $C_b(H)$ such that $F(\mathbf{1}) = 1$. Then $F \in \mathscr{P}(H)$ if and only if it is tight.*

Proof. If $F \in \mathscr{P}(H)$ then it is tight by Theorem 6.2. Assume conversely that F is tight and let (K_n) be an increasing sequence of compact subsets of H such that $\lim_{n \to \infty} F_{K_n}(\mathbf{1}) = 1$. Obviously the sequence of measures (μ_{K_n}) is increasing and bounded. Set

$$\mu(B) = \lim_{n \to \infty} \mu_{K_n}(B), \quad B \in \mathscr{B}(H).$$

Then μ is a Borel measure on H. In fact if $A_m \uparrow A$ in $\mathscr{B}(H)$, we have

$$\lim_{m\to\infty} \mu(A_m) = \lim_{m\to\infty}\lim_{n\to\infty} \mu_n(A_m)$$

$$= \sup_{m\in\mathbb{N}}\sup_{n\in\mathbb{N}} \mu_n(A_m) = \sup_{n\in\mathbb{N}}\sup_{m\in\mathbb{N}} \mu_n(A_m)$$

$$= \lim_{n\to\infty}\lim_{m\to\infty} \mu_n(A_m) = \lim_{n\to\infty} \mu_n(A) = \mu(A).$$

Moreover, by the assumption we have $\mu(H) = 1$. Finally, for any non-negative $\varphi \in C_b(H)$ we have

$$F(\varphi) \geq F_{K_n}(\varphi) = \int_H \varphi(x)\mu_{K_n}(dx).$$

Letting $n \to \infty$ we find that

$$F(\varphi) \geq \int_H \varphi(x)\mu(dx),$$

and the conclusion follows from Proposition 6.3. $\square$

6.3 The Prokhorov theorem

A sequence $(\mu_k) \subset \mathscr{P}(H)$ is said to be *weakly convergent* to a probability measure μ $(\mu_k \rightharpoonup \mu)$ if we have

$$\lim_{k\to\infty} \int_H \varphi(x)\mu_k(dx) = \int_H \varphi(x)\mu(dx) \quad \text{for all } \varphi \in C_b(H). \qquad (6.12)$$

A subset $\Lambda \in \mathscr{P}(H)$ is said to be *weakly relatively compact* if one can extract from any sequence in Λ a subsequence which is weakly convergent to an element of $\mathscr{P}(H)$.

Notice that if $\mu_k \rightharpoonup \mu$ we do not have necessarily

$$\lim_{n\to\infty} \mu_n(B) = \mu(B) \quad \text{for all } B \in \mathscr{B}(H).$$

Consider in fact the particular case when $H = \mathbb{R}$, and set

$$\mu_k = \delta_{1/k}, \quad k \in \mathbb{N}, \ \mu = \delta_0.$$

Then $\mu_k \rightharpoonup \mu$, $\mu_k(\{0\}) = 0$ but $\mu(\{0\}) = 1$.

A subset $\Lambda \subset \mathscr{P}(H)$ is said to be *tight* if there exists an increasing sequence (K_n) of compact sets of H such that

$$\lim_{n\to\infty} \mu(K_n) = 1 \quad \text{uniformly on } \Lambda,$$

or, equivalently, if for any $\varepsilon > 0$ there exists a compact set K_ε such that

$$\mu(K_\varepsilon) \geq 1 - \varepsilon, \quad \mu \in \Lambda.$$

We are now ready to prove the *Prokhorov theorem*.

Theorem 6.7 *Assume that $\Lambda \in \mathscr{P}(H)$ is tight. Then it is weakly relatively compact.* [2]

Proof. Let (K_n) be an increasing sequence of compact sets in H such that

$$\mu(K_n) \geq 1 - \frac{1}{n} \quad \text{for all } \mu \in \Lambda, \ n \in \mathbb{N}. \tag{6.13}$$

For any $\mu \in \Lambda$ we denote by μ_{K_n} the restriction of μ to K_n. For any $n \in \mathbb{N}$, the set of positive functionals $\{F_{\mu_{K_n}} : \mu \in \Lambda\}$ is bounded on $C(K)^*$; consequently, in view of the Banach–Alaoglu theorem (see e.g. H. Brézis [4, Théorème III.15]) it is $*$-relatively compact. Therefore there is a sequence $(\mu_{1,k}) \subset \Lambda$ and a measure $\tilde{\mu}_1$ in K_1 such that

$$\lim_{k \to \infty} \mu_{1,k} = \tilde{\mu}_1 \quad \text{weakly,}$$

and so

$$\lim_{k \to \infty} \int_{K_1} \varphi d\mu_{1,k} = \int_{K_1} \varphi d\tilde{\mu}_1 := F_1(\varphi), \quad \varphi \in C_b(H). \tag{6.14}$$

Analogously there is a subsequence $(\mu_{2,k})$ of $(\mu_{1,k})$ weakly convergent to a measure $\tilde{\mu}_2$ in K_2. Iterating this procedure, we can construct a subsequence $(\mu_{m,k})$ of $(\mu_{m-1,k})$ weakly convergent to a measure $\tilde{\mu}_m$ in K_m for all $m \in N$, and we have

$$\lim_{k \to \infty} \int_{K_m} \varphi d\mu_{m,k} = \int_{K_m} \varphi d\tilde{\mu}_m := F_m(\varphi), \quad \varphi \in C_b(H). \tag{6.15}$$

Since for $\varphi \in C_b(H), \varphi \geq 0$,

$$F_m(\varphi) = \int_{K_{m-1}} \varphi d\mu_m + \int_{K_m \setminus K_{m-1}} \varphi d\mu_m$$

$$= \int_{K_{m-1}} \varphi d\mu_{m-1} + \int_{K_m \setminus K_{m-1}} \varphi d\mu_m \geq F_{m-1}(\varphi).$$

the sequence $(F_n(\varphi)), \varphi \geq 0$ is nondecreasing.

---

[2] The converse is also true, see e.g. K. P. Parthasarathy, [22].

Setting $\nu_k = \mu_{k,k}$, $k \in \mathbb{N}$, we have

$$\lim_{k \to \infty} \int_{K_n} \varphi d\nu_k =: F_n(\varphi), \quad \varphi \in C_b(H), \ n \in \mathbb{N}.$$

Moreover

$$F(\varphi) = \lim_{n \to \infty} F_n(\varphi),$$

is a positive functional. By Theorem 6.6 there is a measure ν such that $F = F_\nu$. It remains to show that $\nu_m \rightharpoonup \nu$.

Let $\varphi \in C_b(H), \varphi \geq 0, \varepsilon > 0, k, n \in \mathbb{N}$. We have

$$\left| \int_H \varphi(x)\nu(dx) - \int_H \varphi(x)\nu_k(dx) \right| = \left| F(\varphi) - \int_H \varphi(x)\nu_k(dx) \right|$$

$$\leq |F(\varphi) - F_n(\varphi)| + \left| F_n(\varphi) - \int_{K_n} \varphi(x)\nu_k(dx) \right| + \int_{K_n^c} \varphi(x)\nu_k(dx)$$

$$\leq |F(\varphi) - F_n(\varphi)| + \left| F_n(\varphi) - \int_{K_n} \varphi(x)\nu_k(dx) \right| + \frac{1}{n}\|\varphi\|_0$$

by (6.7). Finally, let $n_\varepsilon \in \mathbb{N}$ be such that

$$|F(\varphi) - F_{n_\varepsilon}(\varphi)| \leq \frac{\varepsilon}{3}, \quad \frac{1}{n_\varepsilon}\|\varphi\|_0 \leq \frac{\varepsilon}{3},$$

then we have

$$\left| F(\varphi) - \int_H \varphi(x)\nu_k(dx) \right| \leq \frac{2}{3}\varepsilon + \left| F_{n_\varepsilon}(\varphi) - \int_{K_{n_\varepsilon}} \varphi(x)\nu_k(dx) \right|,$$

and the conclusion follows letting $k \to \infty$. $\square$

We conclude this chapter by giving a useful sufficient condition for tightness of a subset Λ of $\mathscr{P}(H)$.

Proposition 6.8 *Let $\Lambda \subset \mathscr{P}(H)$ and let $V: H \to [0, +\infty]$ be a Borel function such that its level sets*

$$K_a: \ = \{x \in H : V(x) \leq a\}, \quad a > 0,$$

are compact for any a sufficiently large. Assume in addition that

$$\sup_{\mu \in \Lambda} \int_H V(x)\mu(dx): \ = \kappa < +\infty. \tag{6.16}$$

Then Λ is tight.

Proof. Let $a > 0$. Then we have

$$\mu(K_a^c) = \int_{\{V > a\}} \mu(dx) \leq \frac{1}{a} \int_H V(x)\mu(dx) \leq \frac{\kappa}{a}, \quad \mu \in \Lambda.$$

Thus

$$\lim_{a \to \infty} \mu(K_a^c) = 0 \quad \text{uniformly on} \quad \mu \in \Lambda.$$

This implies the tightness of Λ. $\square$

Existence and uniqueness of invariant measures

We are given a separable Hilbert space H, (norm $|\cdot|$, inner product $\langle \cdot, \cdot \rangle$) and a Markov semigroup P_t (see Chapter 5 for the definition)

$$P_t \varphi(x) = \int_H \varphi(y) \pi_t(x, dy), \quad t \geq 0, \ \varphi \in C_b(H),$$

on H. We shall assume throughout this chapter that P_t is Feller, that is

$$\varphi \in C_b(H) \implies P_t \varphi \in C_b(H) \quad \text{for all } t \geq 0.$$

In section 7.1 we shall prove the Krylov–Bogoliubov theorem on the existence of invariant measures, whereas section 7.2 is devoted to uniqueness and section 7.3 to application to some stochastic differential equations in $\mathbb{R}^n$.

7.1 The Krylov–Bogoliubov theorem

Assume that for some $x_0 \in H$ there exists the limit

$$\lim_{t \to +\infty} P_t \varphi(x_0) := F_{x_0}(\varphi), \tag{7.1}$$

for all $\varphi \in C_b(H)$. Then it is easy to check that F_{x_0} is a positive functional from $C_b(H)$ into $\mathbb{R}$ such that $F_{x_0}(1) = 1$. Moreover F_{x_0} is *invariant* for P_t in the following sense,

$$F_{x_0}(P_t \varphi) = F_{x_0}(\varphi) \quad \text{for all } \varphi \in C_b(H) \text{ and } t \geq 0. \tag{7.2}$$

In fact, setting in (7.1) $\varphi = P_s \psi$, where $\psi \in C_b(H)$ we find that

$$\lim_{t \to +\infty} P_t P_s \psi(x_0) = F_{x_0}(P_s \psi).$$

On the other hand,

$$\lim_{t \to +\infty} P_t P_s \psi(x_0) = \lim_{t \to +\infty} P_{t+s}\psi(x_0) = F_{x_0}(\psi).$$

Therefore, (7.2) follows.

If in addition $(\pi_t(x_0, \cdot))_{t \geq 0}$ is tight then, by the Prokhorov theorem, there exists $\mu \in \mathscr{P}(H)$ and $t_n \uparrow \infty$ such that $\pi_{t_n}(x_0, \cdot) \rightharpoonup \mu$. Consequently, for all $\varphi \in C_b(H)$ we have

$$P_{t_n}\varphi(x_0) = \int_H \varphi(y)\pi_{t_n}(x_0, dy) \to \int_H \varphi(y)\mu(dy) = F_{x_0}(\varphi),$$

as $n \to \infty$. In other words, $\mu = F_{x_0}$ is invariant for P_t.

We notice that the condition (7.1) is too strong in the applications, however, tightness of the family $(\pi_t(x_0, \cdot))_{t \geq 0}$ only, [1] ensures existence of an invariant measure for P_t. This is a consequence of the following *Krylov–Bogoliubov theorem*.

To formulate the theorem, let us introduce the following notation. For any $T > 0$ we set

$$\mu_T(E) = \frac{1}{T} \int_0^T \pi_t(x_0, E)dt, \quad E \in \mathscr{B}(H), \ T > 0. \tag{7.3}$$

Notice that the integral above is meaningful since the mapping

$$[0, +\infty) \to \mathbb{R}, \ t \mapsto \pi_t(x_0, E) = P_t \mathbf{1}_E(x_0),$$

is Borel in view of Definition 5.1(iii).

Theorem 7.1 *Let P_t be a Markov Feller semigroup. Assume that for some $x_0 \in H$ the set $(\mu_T)_{T>0}$, defined by (7.3), is tight. Then there is an invariant measure for P_t.*

Proof. By the Prokhorov theorem there exists a sequence $T_n \uparrow \infty$ and a probability measure $\mu \in \mathscr{P}(H)$ such that

$$\lim_{n \to \infty} \int_H \varphi(x)\mu_{T_n}(dx) = \int_H \varphi(x)\mu(dx) \quad \text{for all } \varphi \in C_b(H),$$

which, in view of the Fubini theorem, is equivalent to

$$\lim_{n \to \infty} \frac{1}{T_n} \int_0^{T_n} P_t\varphi(x_0)dt = \int_H \varphi(x)\mu(dx) \quad \text{for all } \varphi \in C_b(H). \tag{7.4}$$

Setting $\varphi = P_s\psi$, for $s \geq 0$, we have that $\varphi \in C_b(H)$ since P_t is Feller, and

---

[1] Even a weaker condition, see Remark 7.2 below.

$$\lim_{n\to\infty} \frac{1}{T_n} \int_0^{T_n} P_{t+s}\psi(x_0)dt = \int_H P_s\psi d\mu, \quad \text{for all } \psi \in C_b(H). \quad (7.5)$$

Let us prove that the left-hand side of (7.5) is equal to $\int_H \psi d\mu$. This will prove that μ is invariant. We have in fact, taking into account (7.4),

$$\frac{1}{T_n} \int_0^{T_n} P_{t+s}\psi(x_0)dt = \frac{1}{T_n} \int_s^{T_n+s} P_t\psi(x_0)dt$$

$$= \frac{1}{T_n} \int_0^{T_n} P_t\psi(x_0)dt + \frac{1}{T_n} \int_{T_n}^{T_n+s} P_t\psi(x_0)dt - \frac{1}{T_n} \int_0^s P_t\psi(x_0)dt$$

$$\to \int_H \psi(x)\mu(dx) \text{ as } n \to \infty.$$

□

Remark 7.2 Obviously if the set $(\pi_t(x_0, \cdot)_{t>0})$, is tight then the assumption of Theorem 7.1 is fulfilled and so there exists an invariant measure for P_t.

7.2 Uniqueness of invariant measures

In this section we first introduce the notion of regularity of a Markov semigroup P_t and show that if P_t is regular it possesses at most one invariant measure. Then we prove that if P_t is irreducible and strong Feller it is regular.

Definition 7.3 P_t is said to be regular if for any $t > 0$ all probability measures $\pi_t(x, \cdot)$, $x \in H$, are mutually equivalent.

Proposition 7.4 Assume that the Markov semigroup P_t is regular and possesses an invariant measure μ. Then μ is equivalent to $\pi_t(x, \cdot)$ for all $t > 0$ and all $x \in H$. Moreover, μ is the unique invariant measure of P_t.

Proof. Assume that μ is an invariant measure for P_t. Then for any $A \in \mathscr{B}(H)$ and any $t > 0$ we have

$$\mu(A) = \int_H 1_A d\mu = \int_H P_t 1_A d\mu = \int_H \pi_t(y, A)\mu(dy). \quad (7.6)$$

Let $x \in H$. We claim that μ is equivalent to $\pi_t(x, \cdot)$. In fact if for some $A \in \mathscr{B}(H)$ we have $\pi_t(x, A) = 0$ then $\pi_t(y, A) = 0$ for all $y \in H$ by the assumption of regularity of P_t. Therefore $\mu(A) = 0$ by (7.6) and so $\mu \ll \pi_t(x, \cdot)$.

Conversely if $\mu(A) = 0$ we have, again by (7.6), that $\pi_t(y, A) = 0$ for μ-almost all $y \in H$. Again by the regularity of P_t we conclude that $\pi_t(y, A) = 0$ for all $y \in H$. Therefore $\pi_t(x, \cdot) \ll \mu$.

It remains to prove the uniqueness of μ. Assume by contradiction that there exist two ergodic measures μ and ν. Then μ and ν are singular by Theorem 5.19. Let $A, B \in \mathscr{B}(H)$ be disjoint and such that μ (resp. ν) is concentrated on A (resp. B). Since $\mu(A) = \nu(B) = 1$ we have, by the first part of the proof,

$$\pi_t(x, A) = \pi_t(x, B) = 1, \quad t > 0, \ x \in H,$$

which implies $\pi_t(x, A \cup B) = 2$ for all $t > 0$, $x \in H$, a contradiction. $\square$

We now prove a result due to Khas'minskii, see e.g. [10].

Proposition 7.5 *Assume that the Markov semigroup P_t is strong Feller and irreducible. Then it is regular.*

Proof. Let $t > 0$ and let $x_0 \in H$ be fixed but arbitrary. We have to prove that $\pi_t(x, \cdot)$ is equivalent to $\pi_t(x_0, \cdot)$ for all $x \in H$. For this it is enough to show that if $E \in \mathscr{B}(H)$ is such that $\pi_t(x_0, E) > 0$, we have $\pi_t(x, E) > 0$ for all $x \in H$.

Assume that $\pi_t(x_0, E) > 0$. Then, since for any $h \in (0, t)$ we have by (5.3),

$$\pi_t(x_0, E) = \int_H \pi_h(x_0, dy) \pi_{t-h}(y, E) > 0,$$

there exists $y_0 \in H$ such that $\pi_{t-h}(y_0, E) > 0$. Since P_t is strong Feller, the function $y \mapsto \pi_{t-h}(y, E)$ is continuous, so that there exists $r > 0$ such that

$$\pi_{t-h}(y, E) > 0 \quad \text{for all } y \in B(y_0, r).$$

Let now $x \in H$ be arbitrary, then we have

$$\pi_t(x, E) = \int_H \pi_h(x, dy) \pi_{t-h}(y, E) \geq \int_{B(y_0, r)} \pi_h(x, dy) \pi_{t-h}(y, E) > 0,$$

because $\pi_h(x, B(y_0, r)) > 0$ by the irreducibility of P_t. So, $\pi_t(x, E) > 0$ and we have proved that $\pi_t(x, \cdot)$ is equivalent to $\pi_t(x_0, \cdot)$ as required. $\square$

By Propositions 7.4 and 7.5 we obtain a sufficient condition for the uniqueness of invariant measure of P_t.

Theorem 7.6 *If P_t is strong Feller and irreducible, then it possesses at most one invariant measure.*

We end this section by proving an interesting property of invariant measures when P_t is strong Feller. Let us recall that the *support* of a probability measure $\mu \in \mathscr{P}(H)$ is defined as the intersection of all closed subsets of H having probability 1.

Lemma 7.7 *Let $\mu \in \mathscr{P}(H)$ and let K_μ be the support of μ. Then we have*

$$K_\mu := \{x \in H : \mu(B(x,r)) > 0 \quad \text{for all } r > 0\}. \tag{7.7}$$

Moreover, if $\varphi \in C_b^+(H)$ is such that

$$\int_H \varphi(x)\mu(dx) = 0, \tag{7.8}$$

we have $\varphi(x) = 0$ for all $x \in K_\mu$.

Proof. Assume first that $x_0 \in K_\mu$ and, by contradiction, that there is $r > 0$ such that $\mu(B(x_0,r)) = 0$. This implies that the support of μ does not include x_0, a contradiction.

Conversely, if $x_0 \notin K_\mu$ it is clear that there is $r_0 > 0$ such that $\mu(B(x_0,r_0)) = 0$. So, (7.7) follows.

Finally, let $\varphi \in C_b^+(H)$ fulfill (7.8). Then φ vanishes μ-almost every-where. We want to show that φ vanishes identically on K_μ. Let $x_0 \in H$ such that $\varphi(x_0) > 0$ and assume by contradiction that $x_0 \in K_\mu$. By the continuity of φ it follows that there exists $r_0 > 0$ such that $\varphi(x_0) > 0$ on the ball $B(x_0,r_0)$. But this implies that $\mu(B(x_0,r_0)) = 0$, a contradiction. $\square$

The following result was stated in [17].

Proposition 7.8 *Assume that P_t is strong Feller and that μ and ν are ergodic invariant measures for P_t with supports K_μ and K_ν respectively. Then $K_\mu \cap K_\nu = \varnothing$.*

Proof. We know by Theorem 5.19 that μ and ν are singular. Let $A, B \in \mathscr{B}(H)$ such that $A \cap B = \varnothing$ and $\mu(A) = \nu(B) = 1$. Then for any $t > 0$ we have

$$0 = \mu(A^c) = \int_H \pi_t(x, A^c)\mu(dx) = \nu(B^c) = \int_H \pi_t(x, B^c)\nu(dx).$$

Since P_t is strong Feller the function $x \mapsto \pi_t(x, A^c)$ is continuous. So, by Lemma 7.7, we have

$$\pi_t(x, A^c) = 0 \quad \text{for all } x \in K_\mu.$$

In a similar way we have

$$\pi_t(x, B^c) = 0 \quad \text{for all } x \in K_\nu.$$

Assume now by contradiction that there is $x_0 \in K_\mu \cap K_\nu$. Then we have

$$\pi_t(x_0, A) = \pi_t(x_0, B) = 1,$$

which implies that $\pi_t(x_0, A \cup B) = 2$, a contradiction. $\square$

Remark 7.9 For an interesting generalization of strong Feller semi-groups (asymptotically strong Feller) see [17].

7.3 Application to stochastic differential equations

We consider here the stochastic differential equation

$$X(t) = x + \int_0^t b(X(s))ds + \sqrt{C}\, B(t), \quad t \geq 0, \tag{7.9}$$

where $C \in L(H)$ is symmetric and non-negative, $b \colon \mathbb{R}^n \to \mathbb{R}^n$ fulfills Hypothesis 4.23 and $B(t)$ is a Brownian motion in a probability space $(\Omega, \mathscr{F}, \mathbb{P})$ with values in $H = \mathbb{R}^n$. We denote as before by $X(t, x)$ the solution of (7.9) and set

$$\pi_t(x, \cdot) = X(t, x)_{\#}\mathbb{P}, \quad t \geq 0, \ x \in H,$$

$$P_t\varphi(x) = \mathbb{E}[\varphi(X(t, x))], \quad \varphi \in C_b(\mathbb{R}^n), \ x \in \mathbb{R}^n, \ t \geq 0. \tag{7.10}$$

Moreover, Λ will represent the set (possibly empty) of all invariant measures of P_t.

In subsection 7.3.1 we study existence and in subsection 7.3.3 uniqueness of invariant measures. Finally, subsection 7.3.2 is devoted to the special case when b is decreasing. In this case one can prove both existence and uniqueness of the invariant measure without using the Krylov–Bogoliubov theorem (Theorem 7.1) and the strong Feller property and irreducibility of the transition semigroup (Theorem 7.6).

7.3.1 Existence of invariant measures

A useful tool for proving existence of invariant measures is provided by some auxiliary mappings called *Lyapunov functions*.

Proposition 7.10 *Let* $V : H \longrightarrow [0, +\infty]$ *be a Borel function whose level sets*

$$K_a := \{x \in H : V(x) \leq a\}, \quad a > 0,$$

are compact for any $a > 0$. *Assume that there exists* $x_0 \in \mathbb{R}^n$ *and* $C(x_0) > 0$ *such that*

$$\mathbb{E}[V(X(t, x_0))] \leq C(x_0) \quad \text{for all } t \geq 0. \tag{7.11}$$

Then Λ *is not empty.*

If in addition there exists $C > 0$ *such that*

$$\mathbb{E}[V(X(t, x))] \leq C \quad \text{for all } t \geq 0, x \in H, \tag{7.12}$$

then Λ *is tight and*

$$\int_{\mathbb{R}^n} V(x)\mu(dx) \leq C \quad \text{for all } \mu \in \Lambda. \tag{7.13}$$

Proof. Assume that (7.11) holds. Then for all $a > 0$ we have

$$\pi_t(x_0, K_a^c) = \int_{\{V > a\}} \pi_t(x_0, dy) \leq \frac{1}{a} \int_H V(y)\pi_t(x_0, dy)$$

$$= \frac{1}{a} \mathbb{E}[V(X(t, x_0))] \leq \frac{C(x_0)}{a}.$$

Therefore, the family $(\pi_t(x_0, \cdot))_{t>0}$ is tight, so that there exists an invariant measure thanks to the Krylov–Bogoliubov theorem.

Assume now that (7.12) holds. Let $\mu \in \Lambda$ (we know that Λ is non-empty by the previous step of the proof) and let $\varepsilon > 0$, $t > 0$. Then, by the invariance of μ, we have

$$\int_{\mathbb{R}^n} \frac{V(x)}{1 + \varepsilon V(x)} \mu(dx) = \int_{\mathbb{R}^n} P_t \left(\frac{V}{1 + \varepsilon V} \right) (x)\mu(dx)$$

$$= \int_{\mathbb{R}^n} \mathbb{E}\left[\frac{V(X(t, x))}{1 + \varepsilon V(X(t, x))} \right] \mu(dx) \leq \int_{\mathbb{R}^n} \mathbb{E}[V(X(t, x))]\mu(dx) \leq C.$$

Letting ε tend to 0 yields (7.13). Finally, by Proposition 6.8 it follows that Λ is tight. The proof is complete. $\square$

Remark 7.11 Assume that the assumptions of Proposition 7.10 are fulfilled. Then, in view of the Krein–Milman theorem (see e.g. [26]), the set of all extremal points of Λ is weakly compact and non-empty. Consequently, there exists an invariant ergodic measure thanks to Theorem 5.18.

Example 7.12 Assume that Hypothesis 4.23 is fulfilled and that there exists $\kappa \geq 0$ and $a > 0$ such that [2]

$$\langle b(x), x \rangle \leq \kappa - a|x|^4 \quad \text{for all } x \in \mathbb{R}^n.$$

By the Itô formula, see (4.53), we have

$$\frac{d}{dt}\, \mathbb{E}|X(t,x)|^2 = \mathbb{E}(\langle b(X(t,x)), X(t,x)\rangle) + \text{Tr } C$$

$$\leq \text{Tr } C + 2\kappa - 2a\mathbb{E}|X(t,x)|^2.$$

Consequently,

$$\mathbb{E}|X(t,x)|^2 \leq |x|^2 + \frac{\text{Tr } C + 2\kappa}{2a}, \quad x \in H.$$

So, (7.11) is fulfilled with $V(x) = |x|^2$ and, by Proposition 7.10 it follows that Λ is non-empty. Moreover, again by the Itô formula, we have

$$\frac{d}{dt}\, \mathbb{E}|X(t,x)|^2 \leq \text{Tr } C + 2\kappa - 2a\mathbb{E}|X(t,x)|^4$$

$$\leq \text{Tr } C + 2\kappa - 2a\left(\mathbb{E}|X(t,x)|^2\right)^2.$$

By an elementary comparison result we find that

$$\mathbb{E}|X(t,x)|^2 \leq \frac{\alpha}{\beta}\, \frac{\beta|x|^2(1 + e^{-\frac{t}{2\alpha\beta}}) + \alpha(1 - e^{-\frac{t}{2\alpha\beta}})}{\beta|x|^2(1 - e^{-\frac{t}{2\alpha\beta}}) + \alpha(1 + e^{-\frac{t}{2\alpha\beta}})}$$

$$\leq \frac{\alpha}{\beta}\, \frac{1 + e^{-\frac{t}{2\alpha\beta}}}{1 - e^{-\frac{t}{2\alpha\beta}}}, \quad t \geq 0,\ x \in \mathbb{R}^n,$$

where

$$\alpha = \text{Tr } C + 2\kappa, \quad \beta = 2a.$$

Consequently, (7.12) is fulfilled as well with $V(x) = |x|^2$ and, by Proposition 7.10, it follows that Λ is tight and it contains ergodic invariant measures.

---

[2] The same proof holds with $2 + \epsilon$ replacing 4 for some $\epsilon > 0$. For a generalization, using suitable Lyapunov functions, see [20].

7.3.2 Existence and uniqueness of invariant measures by monotonicity

In this subsection we shall assume that

Hypothesis 7.13
(i) b is locally Lipschitz continuous.
(ii) There exists $\beta > 0$ such that

$$\langle b(x) - b(y), x - y \rangle \leq -\beta |x - y|^2, \quad x, y \in \mathbb{R}^n. \tag{7.14}$$

Condition (ii) means that b is strictly decreasing. We notice that condition (i) can be replaced by the continuity of b, see [10].

Exercise 7.14 Prove that Hypothesis 7.13 implies Hypothesis 4.23.

It is useful to consider problem (4.4) with a negative initial time s (see [10]), that is

$$\begin{cases} dX(t) = b(X(t))dt + \sqrt{C}\, d\overline{B}(t), & t \geq -s \\ \\ X(-s) = x. \end{cases} \tag{7.15}$$

Here $\overline{B}(t)$ is defined for all $t \in \mathbb{R}$ as follows. We take another Brownian motion $B_1(t)$ independent of $B(t)$ and set

$$\overline{B}(t) = \begin{cases} B(t) & \text{if } t \geq 0, \\ \\ B_1(-t) & \text{if } t \leq 0. \end{cases}$$

Now Proposition 4.3 can be generalized in a straightforward way to solve problem (7.15). We denote by $X(t, -s, x)$ its unique solution, that is the solution of the integral equation

$$X(t, -s, x) = x + \int_{-s}^{t} b(X(u, -s, x))du + \sqrt{C}\, (\overline{B}(t) - \overline{B}(-s)).$$

By Proposition 4.7 the law of $X(0, -t, x)$ coincides with that of $X(t, x)$.

We are going to show, following [10], that the law $\pi_t(x, \cdot)$ of $X(t, x)$ is weakly convergent as $s \to +\infty$ to the unique invariant measure ν of P_t. In fact we shall prove a stronger result, namely that there exists the limit

$$\lim_{s \to +\infty} X(0, -s, x) := \eta \quad \text{in } L^2(\Omega, \mathscr{F}, \mathbb{P}; \mathbb{R}^n).$$

The law $\eta_{\#}\mathbb{P}$ of η will be the required invariant measure of P_t.

Proposition 7.15 *Assume that (7.14) holds. Then there exists a square integrable function* $\eta : \Omega \to \mathbb{R}^n$ *such that for any* $x \in \mathbb{R}^n$ *we have*

$$\lim_{s \to +\infty} X(0, -s, x) = \eta \quad \text{in } L^2(\Omega, \mathscr{F}, \mathbb{P}; \mathbb{R}^n). \tag{7.16}$$

Moreover, there exists $c_1 > 0$ *such that*

$$\mathbb{E}(|X(0, -s, x) - \eta|^2) \le c_1 e^{-2\beta s} |x|^2, \quad s > 0. \tag{7.17}$$

Proof. Let $x \in \mathbb{R}^n$ be fixed and consider the solution $X(t) = X(t, -s, x)$ of (7.15). By the Itô formula we have

$$\frac{d}{dt} \mathbb{E}(|X(t)|^2) = 2\langle b(X(t)), X(t) \rangle + \operatorname{Tr} C.$$

It follows, taking into account (7.14), that

$$\frac{d}{dt} \mathbb{E}(|X(t)|^2) = 2\langle b(X(t)) - b(0), X(t) \rangle + 2\langle b(0), X(t) \rangle + \operatorname{Tr} C$$

$$\le -2\beta |X(t)|^2 + 2|b(0)| \, |X(t)| + \operatorname{Tr} C$$

$$\le -\beta |X(t)|^2 + \frac{4}{\beta} |b(0)|^2 + \operatorname{Tr} C.$$

Consequently, setting

$$\gamma := \frac{4}{\beta} |b(0)|^2 + \operatorname{Tr} C,$$

we have

$$\mathbb{E}(|X(t, -s, x)|^2) \le e^{-\beta(t+s)} |x|^2 + \frac{\gamma}{\beta}, \quad t \ge s. \tag{7.18}$$

We now proceed in two steps.

Step 1. There exists $\eta_x \in L^2(\Omega, \mathscr{F}, \mathbb{P}; H)$ such that $\lim_{s \to +\infty} X(0, -s, x) = \eta_x$ in $L^2(\Omega, \mathscr{F}, \mathbb{P}; H)$.

Let $s > s_1$, and set $X_s(t) = X(t, -s, x)$ for all $s > 0$ and $Z(t) = X_s(t) - X_{s_1}(t)$. Then we have

$$\begin{cases} \dfrac{d}{dt} Z(t) = b(X_s(t)) - b(X_{s_1}(t)), & t \ge -s_1, \\[2mm] Z(-s_1) = X_s(-s_1) - x. \end{cases}$$

Taking the scalar product of both sides of the first equation by $Z(t)$ and taking into account (7.14) we obtain

$$\frac{1}{2}\frac{d}{dt}|Z(t)|^2 \le -\beta|Z(t)|^2, \quad t \ge 0,$$

so that

$$|X_s(t) - X_{s_1}(t)|^2 = |Z(t)|^2 \le e^{-2\beta(t+s)}|X_s(-s_1) - x|^2.$$

Recalling (7.18), we see that there exists a constant $c_1 > 0$ such that

$$\mathbb{E}\left(|X_s(0) - X_{s_1}(0)|^2\right) \le c_1 e^{-2\beta s}|x|^2, \quad s > s_1. \tag{7.19}$$

Consequently $(X_s(0))$ is a Cauchy sequence in $L^2(\Omega, \mathscr{F}, \mathbb{P}; H)$. Moreover, (7.17) follows, letting s_1 tend to infinity in (7.19). Step 1 is proved.

Step 2. η_x is independent of x.

Let $x, y \in H$, and set

$$\rho_s(t) = X(t, -s, x) - X(t, -s, y).$$

Then we have

$$\begin{cases} \dfrac{d}{dt}\rho_s(t) = b(X(t, -s, x)) - b(X(t, -s, y)) \\ \rho(-s) = x - y. \end{cases}$$

Taking the scalar product of the first equation by $\rho_s(t)$ and arguing as before, it follows that

$$|\rho_s(t)| \le e^{-\beta(t+s)}|x - y|, \quad t \ge -s.$$

Therefore, as $s \to +\infty$ we obtain $\eta_x = \eta_y$ as required.
The proof is complete. $\square$
We can prove finally that there exists a unique invariant measure for P_t which is in addition strongly mixing.

Theorem 7.16 *Assume that Hypothesis 7.13 holds and let ν be the law of the random variable η defined by (7.16). Then the following statements hold.*

(i) We have

$$\lim_{t \to +\infty} P_t\varphi(x) = \int_H \varphi(y)\nu(dy), \quad x \in H, \; \varphi \in C_b(\mathbb{R}^n). \qquad (7.20)$$

(ii) ν is the unique invariant measure for P_t.

(iii) For any Borel probability measure $\lambda \in \mathscr{P}(H)$ we have

$$\lim_{t \to +\infty} \int_{\mathbb{R}^n} P_t\varphi(x)\lambda(dx) = \int_H \varphi(x)\nu(dx), \quad \varphi \in C_b(\mathbb{R}^n).$$

(iv) There exists $c > 0$ such that for any function $\varphi \in C_b^1(H)$ we have

$$\left| P_t\varphi(x) - \int_{\mathbb{R}^n} \varphi(y)\nu(dy) \right| \leq c\|\varphi\|_1 e^{-\beta t}|x|^2, \quad t \geq 0. \qquad (7.21)$$

Condition (i) expresses the fact that ν is strongly mixing and (7.21) yields exponential convergence to equilibrium of P_t.

Proof. (i) If $\varphi \in C_b(\mathbb{R}^n)$ we have

$$P_t\varphi(x) = \mathbb{E}[\varphi(X(t,x))] = \mathbb{E}[\varphi(X(0,-t,x))].$$

Letting t tend to $+\infty$ yields

$$\lim_{t \to +\infty} P_t\varphi(x) = \mathbb{E}[\varphi(\eta)] = \int_{\mathbb{R}^n} \varphi(y)\nu(dy),$$

because the law of η is precisely ν.

(ii) For any $t, s > 0$ and any $\varphi \in C_b(\mathbb{R}^n)$, we have

$$\int_{\mathbb{R}^n} P_{t+s}\varphi(x)\nu(dx) = \int_{\mathbb{R}^n} P_t P_s\varphi(x)\nu(dx).$$

Letting t tend to $+\infty$ we find, by (7.20),

$$\int_{\mathbb{R}^n} \varphi(y)\nu(dy) = \int_{\mathbb{R}^n} P_s\varphi(y)\nu(dy),$$

so that ν is invariant.

Let us prove uniqueness of ν. Let λ be an invariant probability measure for P_t, that is

$$\int_{\mathbb{R}^n} P_t\varphi(x)\lambda(dx) = \int_{\mathbb{R}^n} \varphi(x)\lambda(dx), \quad \varphi \in C_b(\mathbb{R}^n).$$

Then, letting t tend to $+\infty$ yields, thanks to (7.20),

$$\int_{\mathbb{R}^n} \varphi(x)\nu(dx) = \int_{\mathbb{R}^n} \varphi(x)\lambda(dx), \quad \varphi \in C_b(\mathbb{R}^n),$$

which implies that $\lambda = \nu$.

(iii) follows again by (7.20). Let us prove finally (iv). Since

$$P_t\varphi(x) - \int_{\mathbb{R}^n} \varphi(y)\nu(dy) = \mathbb{E}(\varphi(X(0,-t,x)) - \varphi(\eta)),$$

then, taking into account (7.17), we have

$$\left| P_t\varphi(x) - \int_{\mathbb{R}^n} \varphi(y)\nu(dy) \right| \leq \|\varphi\|_1 \mathbb{E}|\varphi(X(0,-t,x)) - \varphi(\eta)|$$

$$\leq \|\varphi\|_1 \sqrt{c_1} e^{-\beta t}|x|^2, \quad t \geq 0.$$

$\square$

7.3.3 Uniqueness of invariant measures

We assume here, besides Hypothesis 4.23, that the transition semigroup P_t has at least an invariant measure. To prove uniqueness we shall use Theorem 7.6, more precisely we shall present sufficient conditions for the irreducibility and strong Feller properties of P_t.

Let us first prove irreducibility. To this purpose we shall use a simple result on controllability of the deterministic problem,

$$u(t) = x + \int_0^t b(u(s))ds + \sigma(t), \quad t \geq 0 \tag{7.22}$$

(here σ is the *control*). Then we will compare (7.22) with the stochastic integral equation,

$$X(t,x) = x + \int_0^t b(u(X(s,x)))ds + B(t), \quad t \geq 0. \tag{7.23}$$

Lemma 7.17 *Assume that b is Lipschitz continuous and that $C = I$. Let $x, z \in H, R > 0$, and $T > 0$ be fixed. Then there exists $\sigma \in C([0,T]; H)$ such that $u(0) = x, u(T) = z$.*

Proof. It is enough to set

$$u(t) = \frac{T-t}{T} x + \frac{t}{T} z, \quad t \in [0,T],$$

and

$$\sigma(t) = u(t) - x - \int_0^t b(u(s))ds, \quad t \in [0,T].$$

$\square$

Proposition 7.18 P_t *is irreducible.*

Proof. Let $x, z \in H, R > 0$, and $T > 0$ be fixed. We have to prove that

$$\mathbb{P}(|X(T, x) - z| < R) > 0.$$

By Lemma 7.17 there exists $\sigma \in C([0, T]; H)$ such that $u(0) = x, u(T) = z$, where u is the solution of (7.22). Now by subtracting (7.23) and (7.22) we have

$$X(t, x) - u(t) = \int_0^t [b(X(s, x)) - b(u(s))]ds + B(t) - \sigma(t).$$

Therefore

$$|X(t, x) - u(t)| \leq M \int_0^t |X(s, x) - u(s)|ds + |B(t) - \sigma(t)|, \quad (7.24)$$

where M is the Lipschitz constant of b. By the Gronwall lemma it follows that

$$|X(T, x) - z| \leq \int_0^T e^{(T-s)M}|B(s) - \sigma(s)|ds.$$

Now by the Hölder inequality we have

$$|X(T, x) - z| \leq \sqrt{T}\, e^{TM}\|B - \sigma\|_{L^2(0,T;H)}.$$

Consequently, we have

$$\mathbb{P}\left(|X(T, x) - z| \geq r\right) \leq \mathbb{P}\left(\|B - \sigma\|_{L^2(0,T;H)} \geq \frac{r}{\sqrt{T}}\, e^{-TM}\right) < 1,$$

since B is a Gaussian random variable on $L^2(0, T; \mathbb{R}^n)$ with nondegenerate covariance, by Proposition 3.14.

In conclusion we have

$$\mathbb{P}\left(|X(T, x) - z| < r\right) > 0 \quad \text{for all } r > 0,$$

and irreducibility of P_t is proved. $\square$

We finally prove that P_t is strong Feller.

Proposition 7.19 *Assume, besides Hypothesis* 4.23, *that* $C = I$ *and that there exists* $N > 0$ *such that*

$$\langle b(x) - b(y), x - y \rangle \leq N|x - y|^2, \quad x, y \in \mathbb{R}^n. \quad (7.25)$$

Then P_t *is strong Feller.*

Proof. Step 1. For all $\varphi \in C_b^3(\mathbb{R}^n)$ we have

$$|D_x P_t \varphi(x)| \leq t^{-1/2} e^{tN} \|\varphi\|_0, \quad x \in \mathbb{R}^n. \tag{7.26}$$

Set $u(t, x) = P_t \varphi(x)$. Then we know by Theorem 4.19 that u is the strict solution of the Cauchy problem

$$\begin{cases} D_t u = \dfrac{1}{2} \Delta u + \langle b(x), D_x u \rangle \\[2mm] u(0, x) = \varphi(x), \quad x \in \mathbb{R}^n. \end{cases} \tag{7.27}$$

To prove estimate (7.26) we use a classical argument due to Bernstein. We set

$$z(t, x) = u^2(t, x) + \frac{t}{2} |D_x u(t, x)|^2, \quad t \geq 0, \, x \in H,$$

and look for an equation fulfilled by z. For this we compute successively $D_t(u^2)$ and $D_t(|D_x u|^2)$. We find, after elementary computations,

$$D_t(u^2) = \frac{1}{2} \Delta(u^2) - \frac{1}{2} |D_x u|^2 + \langle b, D_x(u^2) \rangle \tag{7.28}$$

and

$$D_t(|D_x u|^2) = \frac{1}{2} \Delta(|D_x u|^2) - \frac{1}{2} \sum_{i,j=1}^n |D_{x_i} D_{x_j} u|^2$$

$$+ \langle b, D_x(|D_x u|^2) \rangle + \langle D_x b \cdot D_x u, D_x u \rangle. \tag{7.29}$$

By (7.28) and (7.29) it follows that

$$D_t z = \frac{1}{2} \Delta z - \frac{1}{2} \sum_{i,j=1}^n |D_{x_i} D_{x_y} u|^2 + \langle b, D_x z \rangle + 2t \langle D_x b \cdot D_x u, D_x u \rangle$$

$$\leq \frac{1}{2} \Delta z + \langle b, D_x z \rangle + 2N z. \tag{7.30}$$

Since $z(0, \cdot) = \varphi^2$ we have

$$z(t, \cdot) \leq P_t(\varphi^2) + 2N \int_0^t P_{t-s} z(s, \cdot) ds,$$

and therefore

$$\|z(t, \cdot)\|_0 \leq \|\varphi\|_0^2 + 2N \int_0^t \|z(s, \cdot)\|_0 ds.$$

By the Gronwall lemma it follows that

$$|z(t,x)| \le e^{Nt}\|\varphi\|_0, \quad t \ge 0, \ x \in \mathbb{R}^n,$$

whiich yields

$$|D_x u(t,x)| \le t^{-1/2} e^{tN}\|\varphi\|_0, \quad t \ge 0, \ x \in \mathbb{R}^n,$$

and (7.26) is proved when $\varphi \in C_b^3(H)$.

Step 2. For all $\varphi \in C_b(\mathbb{R}^n)$, (7.26) holds.

Let $(\varphi_n) \subset C_b^3(H)$ such that $\varphi_n \to \varphi$ in $C_b(H)$. Set

$$u_n(t,x) = P_t\varphi_n(x), \quad x \in H, \ t \ge 0.$$

Then by (7.30) it follows that, for any $m,n \in \mathbb{N}$,

$$|Du_n(t,x) - Du_m(t,x)| \le t^{-1/2} e^{tM}\|\varphi_n - \varphi_m\|_0.$$

This implies that $u(t,\cdot) \in C_b^1(\mathbb{R}^n)$ for all $t > 0$ and that (7.26) holds.

Step 3. Conclusion.

Fix $t > 0$, let $x, y \in \mathbb{R}^n$ and let $\varphi \in B_b(\mathbb{R}^n)$. Let us consider the (signed) measure $\zeta_{x,y} = \pi_t(x,\cdot) - \pi_t(y,\cdot)$, and choose a sequence $(\varphi_n) \subset C_b^1(\mathbb{R}^n)$ such that

(i) $\lim\limits_{n\to\infty} \varphi_n(z) = \varphi(z)$ for $\zeta_{x,y}$-almost $z \in \mathbb{R}^n$.

(ii) $\|\varphi_n\|_0 \le \|\varphi\|_0, \quad n \in \mathbb{N}.$

Then we have

$$P_t\varphi(x) - P_t\varphi(y) = \int_H \varphi(z)\zeta_{x,y}(dx)$$

and so, by the dominated convergence theorem,

$$P_t\varphi(x) - P_t\varphi(y) = \lim_{n\to\infty} \int_H \varphi_n(z)\zeta_{x,y}(dx).$$

Now, using (7.26), we conclude that the inequality

$$|P_t\varphi(x) - P_t\varphi(y)| \le t^{-1/2}\, e^{Mt}\, \|\varphi\|_0\, |x - y|$$

holds for all $x, y \in H$. So, $P_t\varphi$ is continuous as required. $\square$

We can now prove a uniqueness result.

Proposition 7.20 *Assume that b is Lipschitz continuous and that $C = I$. Then the transition semigroup P_t has at most one invariant measure.*

Proof. P_t is irreducible in view of Proposition 7.18 and strong Feller by Proposition 7.19. Thus, the conclusion follows from Theorem 7.6. $\square$

Examples of Markov semigroups

8.1 Introduction

In this chapter we present some examples of Markov semigroups in a real separable Hilbert space H. More precisely, in section 8.2 we introduce the *heat semigroup* and in section 8.3 the *Ornstein–Uhlenbeck semigroup*. We shall study several properties, such as irreducibility, strong Feller property and regularity of these semigroups.

For several computations it is useful to introduce the space $\mathscr{E}(H)$ of all *exponential functions*, that is the linear span in $C_b(H)$ of all real and imaginary parts of functions φ_h, $h \in H$, where

$$\varphi_h(x) = e^{i\langle h,x \rangle}, \quad x \in H.$$

The space $\mathscr{E}(H)$ is not dense in $C_b(H)$. However the following approximation result holds.

Lemma 8.1 *For all $\varphi \in C_b(H)$ there exists a two-index sequence $(\varphi_{k,n}) \subset \mathscr{E}(H)$ such that*

$$(i) \quad \lim_{k \to \infty} \lim_{n \to \infty} \varphi_{k,n}(x) = \varphi(x) \quad \textit{for all } x \in H,$$

$$(ii) \quad \|\varphi_{k,n}\|_0 \le \|\varphi\|_0 + \frac{1}{n} \quad \textit{for all } n, k \in \mathbb{N}.$$

(8.1)

Proof. We divide the proof into two steps.

Step 1. The case when H is finite dimensional.

Let $\dim H = N$ and let $\varphi \in C_b(H)$. By the Stone–Weierstrass theorem, for any $R \in \mathbb{N}$ there exists a sequence $(\psi_{R,n}) \subset \mathscr{E}(H)$ such that

$$|\varphi(x) - \psi_{R,n}(x)| \le \frac{1}{n} \quad \text{for all } x \in B_R,$$

where B_R is the ball in H with center 0 and radius R, and

$$\|\psi_{R,n}\|_0 \leq \|\varphi\|_0 + \frac{1}{n} \quad \text{for all } n \in \mathbb{N}.$$

Now the conclusion follows by a standard diagonal extraction argument.

Step 2. The case when dim $H = \infty$.

Given $\varphi \in C_b(H)$, set

$$\varphi_k(x) = \varphi(P_k x), \quad x \in H, \ k \in \mathbb{N}$$

where P_k is defined by $P_k x = \sum_{h=1}^{k} \langle x, e_h \rangle e_h$ and (e_k) is a complete orthonormal system in H. By Step 1, for any $k \in \mathbb{N}$, there exists a sequence $(\varphi_{k,n})_{n \in \mathbb{N}} \subset \mathscr{E}(H)$ such that

$$\lim_{n \to \infty} \varphi_{k,n}(x) = \varphi_k(x), \quad x \in H,$$

and $\|\varphi_{k,n}\|_0 \leq \|\varphi_k\|_0 \leq \|\varphi\|_0 + \frac{1}{n}$. Now, for any $x \in H$ we have

$$\lim_{k \to \infty} \lim_{n \to \infty} \varphi_{k,n}(x) = \lim_{k \to \infty} \varphi_k(x) = \varphi(x) \quad \text{for all } x \in H,$$

and the conclusion follows. $\square$

8.2 The heat semigroup

We are given a linear operator $Q \in L_1^+(H)$, we denote by (e_k) a complete orthonormal system in H and by (λ_k) a sequence of non-negative numbers such that

$$Q e_k = \lambda_k e_k, \quad k \in \mathbb{N}.$$

Let us define

$$U_t \varphi(x) = \int_H \varphi(y) N_{x,tQ}(dy), \quad \varphi \in C_b(H), \ t \geq 0. \tag{8.2}$$

Other equivalent expressions for U_t are, as is easily checked,

$$U_t \varphi(x) = \int_H \varphi(x+y) N_{tQ}(dy)$$

$$= \int_H \varphi(x + \sqrt{t}\, y) N_Q(dy), \ \varphi \in C_b(H), \ t \geq 0. \tag{8.3}$$

Proposition 8.2 U_t *is a Markov semigroup and we have*

$$U_t\varphi(x) = \int_H \varphi(y)\pi_t(x, dy), \quad \varphi \in B_b(H),$$

where

$$\pi_t(x, \cdot) = N_{x,tQ}, \quad t > 0, \ x \in H.$$

Moreover U_t is a strongly continuous semigroup in $C_b(H)$.

Proof. Let us prove the semigroup law,

$$U_{t+s}\varphi = U_t U_s \varphi, \quad t, s \geq 0, \tag{8.4}$$

holds. In view of Lemma 8.1 and the dominated convergence theorem, it is enough to prove (8.4) for any function $\varphi_h(x) = e^{i\langle x,h\rangle}$ with $x, h \in H$. But in this case (8.4) follows immediately (recalling the expression for the Fourier transform of a Gaussian measure (1.5)) from the identity

$$U_t\varphi_h(x) = \int_H e^{i\langle y,h\rangle} N_{x,tQ}(dy) = e^{i\langle h,x\rangle} e^{-\frac{t}{2}\langle Qh,h\rangle}. \tag{8.5}$$

It remains to prove that U_t is strongly continuous. Let $\varphi \in C_b(H)$. Since by (8.3),

$$U_t\varphi(x) = \int_H \varphi(x + \sqrt{t}\, z)N_Q(dz), \quad x \in H,$$

we have

$$|U_t\varphi(x) - \varphi(x)| = \left| \int_H [\varphi(x + \sqrt{t}\, z) - \varphi(x)]N_Q(dz) \right|$$

$$\leq \int_H \omega_\varphi(\sqrt{t}\, z)N_Q(dz),$$

where ω_φ is the uniform continuity modulus of φ. [1] Consequently, for any $R > 0$ we have

$$|U_t\varphi(x) - \varphi(x)| \leq 2\|\varphi\|_0 N_Q(B_R) + \int_{B_R} \omega_\varphi(\sqrt{t}\, z)N_Q(dz),$$

and the conclusion follows letting $R \to \infty$ and $t \to 0$. $\square$

U_t is called the *heat semigroup*. If H has finite dimension d and $Q = I$, it is easy to see that

---

[1] $\omega_\varphi(r) = \sup\{|\varphi(x) - \varphi(y)| : x, y \in H, |x - y| \leq r\}$.

$$U_t\varphi(x) = (2\pi t)^{-n/2} \int_{\mathbb{R}^d} e^{-\frac{1}{2t}|x-y|^2} \varphi(y)dy, \quad \varphi \in C_b(H), \ x \in H. \quad (8.6)$$

Consequently, if $\varphi \in C_b(H)$, $u(t,x) = U_t\varphi(x)$ is the unique classical solution to the heat equation

$$\begin{cases} D_t u(t,x) = \dfrac{1}{2} \Delta u(t,x), & x \in H, \ t > 0, \\ \\ u(0,x) = \varphi(x), & x \in H. \end{cases} \quad (8.7)$$

This means that the mapping $[0,+\infty) \times H \to \mathbb{R}$, $(t,x) \mapsto u(t,x)$ is continuous, its derivatives $u_t, u_{x_i x_j}$, $i,j = 1, \ldots, n$ exist on $(0,+\infty) \times H$ and u fulfills (8.7).

Remark 8.3 Assume that $Q \neq 0$. Then it is easy to see that U_t has no invariant probability measures. Assume in fact by contradiction that there exists $\mu \in \mathscr{P}(H)$ invariant for U_t. Then for any exponential function $\varphi_h(x) = e^{i\langle x,h\rangle}$ we have, taking into account (8.5),

$$\hat{\mu}(h) = \int_H e^{i\langle x,h\rangle}\mu(dx) = \int_H U_t e^{i\langle x,h\rangle}\mu(dx) = e^{-\frac{t}{2}\langle Qh,h\rangle}\hat{\mu}(h),$$

for all $t > 0$, a contradiction.

It is well known that in finite dimensions when Ker $Q = \{0\}$ the heat semigroup (8.6) has a regularizing property in the sense that for any $\varphi \in C_b(H)$ and any $t > 0$, we have $U_t\varphi \in C_b^\infty(H)$.

In infinite dimensions this result is wrong, we can only show that $U_t\varphi$ is regular in the directions of the Cameron–Martin space $Q^{1/2}(H)$.

Proposition 8.4 *Let $\varphi \in C_b(H)$ and $z \in H$. Then there exists the directional derivative*

$$\langle D_x U_t\varphi(x), Q^{1/2}z\rangle = \lim_{h\to 0} \frac{1}{h}\left(U_t\varphi(x + hQ^{1/2}z) - U_t\varphi(x)\right),$$

given by

$$\langle D_x U_t\varphi(x), Q^{1/2}z\rangle = \frac{1}{\sqrt{t}} \int_H \langle h, (tQ)^{-1/2}y\rangle \varphi(x + y)N_{tQ}(dy). \quad (8.8)$$

Proof. Write

$$U_t\varphi(x + hQ^{1/2}z) = \int_H \varphi(x + hQ^{1/2}z + y)N_{tQ}(dy)$$

$$= \int_H \varphi(x + y)N_{hQ^{1/2}z,tQ}(dy).$$

Then, by the Cameron–Martin formula (Theorem 2.8) we have

$$\frac{dN_{hQ^{1/2}z,tQ}(dy)}{dN_{tQ}(dy)}(y) = e^{-\frac{h^2}{2t}|z|^2+\frac{h}{\sqrt{t}}\langle z,(tQ)^{-1/2}y\rangle}, \quad y \in H.$$

It follows that

$$U_t\varphi(x + hQ^{1/2}z) = \int_H \varphi(x + y)e^{-\frac{h^2}{2t}|z|^2+\frac{h}{\sqrt{t}}\langle z,(tQ)^{-1/2}y\rangle} N_{tQ}(dy).$$

Taking the derivative with respect to h and setting $h = 0$ yields (8.8).
$\square$

8.2.1 Initial value problem

We consider now the heat equation,

$$\begin{cases} D_t u(t, x) = \dfrac{1}{2} \operatorname{Tr}[QD_x^2 u(t, x)] = Lu(t, \cdot)(x), \quad t \geq 0, \ x \in H, \\[2mm] u(0, x) = \varphi(x). \end{cases} \tag{8.9}$$

where

$$L\varphi(x) = \frac{1}{2} \operatorname{Tr}[QD_x^2\varphi(x)], \quad \varphi \in C_b^2(H), \quad x \in H.$$

Proposition 8.5 *Let* $\varphi \in C_b^2(H)$ *and set*

$$u(t, x) = U_t\varphi(x), \quad t \geq 0, \ x \in H.$$

Then u is the unique solution of problem (8.9).

Proof. *Existence.* By (8.3) we have

$$D_t u(t, x) = \frac{1}{2\sqrt{t}} \int_H \langle D_x\varphi(x + \sqrt{t}y), z\rangle N_Q(dy). \tag{8.10}$$

Consequently, differentiating (8.8) with $z = e_k$, in the direction e_k, it follows that

$$\langle D_x^2 u(t, x)e_k, e_k\rangle = \frac{1}{\sqrt{\lambda_k t}} \int_H y_k D_{x_k}\varphi(x + \sqrt{t}\,y)N_{tQ}(dy).$$

Therefore

$$\frac{1}{2} \operatorname{Tr}[QD_x^2 u(t, x)] = \sum_{k=1}^{\infty} \langle D_x^2 u(t, x)e_k, e_k\rangle$$

$$= \frac{1}{\sqrt{t}} \int_H \langle y, D_x\varphi(x + \sqrt{t}\,y)\rangle\varphi N_{tQ}(dy).$$

Comparing with (8.10) yields the conclusions.

Uniqueness. Let $v : [0, T] \times H \to \mathbb{R}$ be a continuous function such that

$$D_t v(t, \cdot) = L(t, \cdot), \quad v(0, \cdot) = \varphi.$$

Then we have

$$D_s U_{t-s} v(s, \cdot) = -L U_{t-s} v(s, \cdot) + U_{t-s} v(s, \cdot), \quad s \in [0, t].$$

Since

$$L U_t \psi = U_t L \psi, \quad \psi \in C_b^2(H),$$

we obtain

$$D_s U_{t-s} v(s, \cdot) = 0, \quad s \in [0, t].$$

This implies that $D_s U_{t-s} v(s, \cdot)$ is constant so that $v(s, \cdot) = U_s \varphi$. This proves uniqueness. $\square$

Remark 8.6 One could ask what happens if we consider problem (8.9) with $Q = I$. Obviously definition (8.2) is not meaningful in this case since I is not of trace class. However, one can show that problem (8.9) can still have a solution but only for very special initial data φ, see [10, Proposition 3.1.2].

We study now irreducibility of U_t.

Proposition 8.7 *If* $\mathrm{Ker}\ Q = \{0\}$ *then* U_t *is irreducible.*

Proof. In fact, since the Gaussian measure $N_{x,tQ}$ is nondegenerate for all $x \in H$ and $t > 0$, it is full by Proposition 1.25. Now the conclusion follows from (8.2). $\square$

Let us see whether U_t is strong Feller. If H is finite dimensional it is easy to check that U_t is strong Feller if and only if $\mathrm{Ker}\ Q = \{0\}$. In the infinite dimensional case the strong Feller property never holds. We have in fact the following result, see [11].

Proposition 8.8 *Let* $\mathrm{Ker}\ Q = \{0\}$ *and let* H *be infinite dimensional. Then* U_t *is not strong Feller.*

Proof. Let us choose a subspace $H_0 \subset H$ (with Borel embedding) different from H and such that $Q^{1/2}(H) \subset H_0$ and $N_{tQ}(H_0) = 1$, $t \geq 0$. For this it is enough to take a nondecreasing sequence (α_k) of positive numbers such that $\alpha_k \uparrow +\infty$, and

$$\sum_{k=1}^{\infty} \alpha_k \lambda_k < +\infty,$$

and to define

$$H_0 = \left\{ x \in H : \sum_{k=1}^{\infty} \alpha_k |x_k|^2 < +\infty \right\}.$$

Then $Q^{1/2}(H) \subset H_0$ and $N_{tQ}(H_0) = 1$ since

$$\int_H \sum_{k=1}^{\infty} \alpha_k |x_k|^2 N_{tQ}(dx) = t \sum_{k=1}^{\infty} \alpha_k \lambda_k < +\infty.$$

We show now that

$$U_t \mathbf{1}_{H_0} = \mathbf{1}_{H_0}, \quad t \geq 0. \tag{8.11}$$

This will prove that U_t is not strong Feller because $\mathbf{1}_{H_0}$ is obviously not continuous.

We have in fact

$$U_t \mathbf{1}_{H_0}(x) = \int_{H_0} \mathbf{1}_{H_0}(x + y) N_{tQ}(dy) + \int_{H_0^c} \mathbf{1}_{H_0}(x + y) N_{tQ}(dy).$$

Now (8.11) follows taking into account that for $x \in H_0$, $H_0 + x = H_0$, and for $x \notin H_0$, $(H_0 + x) \cap H_0 = \emptyset$. $\square$

We say that a function φ is *harmonic* if $U_t \varphi = \varphi$, $t \geq 0$. The following generalization of the *Liouville theorem* holds (see [11]).

Proposition 8.9 *Assume that* $\mathrm{Ker}\, Q = \{0\}$ *and that* $\varphi \in C_b(H)$ *is harmonic. Then* φ *is constant.*

Proof. We have in fact from (8.8)

$$|\langle D_x U_t \varphi(x), Q^{1/2} z \rangle| \leq \frac{1}{\sqrt{t}} \|\varphi\|_0. \tag{8.12}$$

Since $U_t \varphi = \varphi$ this implies that

$$\langle D_x \varphi(x), Q^{1/2} z \rangle = 0 \quad \text{for all } z \in H.$$

Consequently φ is constant on the Cameron–Martin space $Q^{1/2}(H)$. Since $Q^{1/2}(H)$ is dense in H and φ is continuous this implies that φ is constant on H. $\square$

8.3 The Ornstein–Uhlenbeck semigroup

Let A be the infinitesimal generator of a strongly continuous semigroup e^{tA} [2] and let $C \in L^+(H)$. We shall assume throughout this section that

---

[2] See Appendix A.

Hypothesis 8.10 *The linear operator Q_t defined by*

$$Q_t x = \int_0^t e^{sA} C e^{sA^*} x \, ds, \quad x \in H,$$

is of trace-class for all $t \geq 0$.

Exercise 8.11 Let $C = I$ and let A be a self-adjoint operator in H such that

$$A e_k = -\alpha_k e_k, \quad k \in \mathbb{N},$$

where (e_k) is a complete orthonormal system in H and $\alpha_k > 0$ for all $k \in \mathbb{N}$.

Find conditions on the sequence (α_k) in order to show that Q_t is of trace class.

For all $t \geq 0$, $x \in H$, define

$$R_t \varphi(x) = \int_H \varphi(y) N_{e^{tA}x, Q_t}(dy)$$

$$\tag{8.13}$$

$$= \int_H \varphi(e^{tA}x + y) N_{Q_t}(dy), \quad \varphi \in B_b(H).$$

Notice that the space of all exponential functions $\mathscr{E}(H)$ is invariant for R_t for all $t \geq 0$. In fact, if $\varphi_h(x) = e^{i\langle h, x \rangle}$, $x \in H$, we have, recalling (1.5),

$$R_t \varphi_h = e^{-\frac{1}{2}\langle Q_t h, h \rangle} \varphi_{e^{tA^*}h}. \tag{8.14}$$

Exercise 8.12 Show that R_t can be defined on all Borel functions φ such that

$$|\varphi(x)| \leq e^{\varepsilon |x|^2}, \quad x \in H,$$

with ε sufficiently small.

Show, in particular, that

$$R_t(|x|^2) = |e^{tA}x|^2 + \operatorname{Tr} Q_t, \quad t \geq 0, \tag{8.15}$$

Proposition 8.13 *Assume that Hypothesis 8.10 holds. Then R_t is a Markov semigroup. Its probability kernel is given by*

$$\pi_t(x, \cdot) = N_{e^{tA}x, Q_t}, \quad x \in H, \ t \geq 0.$$

Proof. In view of Lemma 8.1 it is enough to prove that

$$R_{t+s}\varphi_h = R_t R_s \varphi_h, \quad t, s \geq 0.$$

for all $\varphi_h(x) = e^{i\langle x, h\rangle}$, $h \in H$. In fact by (8.14) it follows that

$$R_s R_t \varphi_h(x) = \exp\left\{i\langle e^{(t+s)A}x, h\rangle - \frac{1}{2}\langle (Q_t + e^{tA}Q_s e^{tA^*})h, h\rangle\right\}.$$

Since

$$Q_t + e^{tA}Q_s e^{tA^*} = Q_t + \int_0^s e^{(t+s)A}C e^{(t+s)A^*}\,ds$$

$$= Q_t + \int_t^{t+s} e^{uA}C e^{uA^*}\,du = Q_{t+s},$$

the conclusion follows. $\square$

Example 8.14 Assume that dim $H = d$, and consider the stochastic differential equation

$$\begin{cases} dX = AX\,dt + \sqrt{C}\,dB(t) \\ X(0) = x, \end{cases} \tag{8.16}$$

where B is a standard d-dimensional Brownian motion and $C \in L^+(H)$. We know by Proposition 4.10 that the problem (8.16) has a unique solution given by the formula

$$X(t, x) = e^{tA}x + \int_0^t e^{(t-s)A}\sqrt{C}\,dB(s)$$

and that $X(t, x)_\#\mathbb{P} = N_{e^{tA}x, Q_t}$. Consequently, the corresponding transition semigroup is given by (8.13) and the Kolmogorov equation reads as follows

$$\begin{cases} D_t u = \frac{1}{2}\,\mathrm{Tr}\,[CD_x^2 u] + \langle Ax, D_x u\rangle \\ u(0, x) = \varphi(x). \end{cases} \tag{8.17}$$

By Proposition 8.7 it follows that R_t is irreducible if and only if Ker $Q_t = \{0\}$. This happens in particular when Ker $C = \{0\}$. In fact if $x_0 \in H$ is such that $Q_t x_0 = 0$ we have

$$0 = \langle Q_t x_0, x_0\rangle = \int_0^t |\sqrt{C}e^{sA}x_0|^2 ds = 0,$$

which implies that $x_0 = 0$.

8.3.1 Smoothing property of the Ornstein–Uhlenbeck semigroup

We assume here that Hypothesis 8.10 holds. We shall prove that the Ornstein–Uhlenbeck semigroup R_t, contrary to the heat semigroup, may have smoothing properties in infinite dimensions. More precisely, it may happen that $R_t\varphi \in C_b^\infty(H)$ for all $\varphi \in B_b(H)$ and all $t > 0$. For this we need the following assumption,

$$e^{tA}(H) \subset Q_t^{1/2}(H), \quad t > 0. \tag{8.18}$$

So, we assume that (8.18) is fulfilled and for any $t > 0$ we set $\Gamma(t) = Q_t^{-1/2}e^{tA}$, where $Q_t^{-1/2}$ is the *pseudo-inverse* of $Q_t^{1/2}$. Let us recall that if $z \in Q^{1/2}(H)$, then by $Q_t^{-1/2}z$ we mean the element of minimal norm of the hyperplane

$$\{v \in H : Q^{1/2}v = z\}.$$

It is easy to see that the operator $\Gamma(t)$ is closable (see subsection A.1.1). Since it is defined in the whole H we have $\Gamma(t) \in L(H)$ for all $t > 0$ by the closed graph theorem, see e.g. [26].

Assumption (8.18) is related to the null controllability of the following deterministic controlled equation in $[0, T]$, see e.g. [11],

$$y'(t) = Ay(t) + \sqrt{C}\, u(t), \quad y(0) = x, \tag{8.19}$$

where $x \in H$ and $u \in L^2(0, T; H)$. Here y represents the *state* and u the *control* of system (8.19). Equation (8.19) has a unique (mild) solution $y(\cdot; u)$, given by the variation of constants formula

$$y(t; u) = e^{tA}x + \int_0^t e^{(t-s)A}\sqrt{C}\, u(s)ds, \quad t \geq 0.$$

We recall that system (8.19) is said to be *null controllable* if for any $T > 0$ there exists $u \in L^2(0, T; H)$ such that $y(T; u) = 0$. One can show, see [27], that system (8.19) is null controllable if and only if the condition (8.18) is fulfilled. In this case, for any $x \in H$, $|\Gamma(t)x|^2$ is the *minimal energy* for driving x to 0, that is

$$|\Gamma(t)x|^2 = \inf\left\{\int_0^T |u(s)|^2 ds : u \in L^2(0, T; U), \ y(T; u) = 0\right\}. \tag{8.20}$$

Remark 8.15 It is important to notice that if $C = I$ (or even if C has a continuous inverse), system (8.19) is always null controllable. In fact, setting $u(t) = -\frac{1}{T} e^{tA}x$, one has $y(T; u) = 0$. Thus in this case (8.18) is fulfilled. Moreover, setting in (8.20)

$$u(t) = -\frac{1}{T} e^{tA}x, \quad t \geq 0,$$

we obtain a bound for the minimal energy, namely

$$|\Gamma(t)x|^2 \leq T^{-2} \int_0^T |e^{sA}x|^2 ds, \quad t > 0, \ x \in H.$$

Now, if $M > 0$ and $\omega \in \mathbb{R}$ are constants such that

$$\|e^{tA}\| \leq M e^{\omega t}, \quad t \geq 0,$$

we obtain

$$\|\Gamma(t)\| \leq \frac{M}{\sqrt{t}} \sup_{s \in [0,t]} e^{2\omega s}, \quad t > 0. \tag{8.21}$$

We can prove now the result,

Theorem 8.16 *Assume that Hypothesis 8.10 holds. Then the following statements are equivalent.*

(i) Condition (8.18) is fulfilled.
(ii) If $\varphi \in B_b(H)$ and $t > 0$ we have $R_t\varphi \in C_b^\infty(H)$.

Proof. $(i) \Rightarrow (ii)$. Let $x \in H$ and $t > 0$. Since by (8.18) $e^{tA}x \in Q_t^{1/2}(H)$, the measures $N_{e^{tA}x, Q_t}$ and N_{Q_t} are equivalent thanks to the Cameron–Martin theorem (Theorem 2.8) and moreover

$$\frac{dN_{e^{tA}x, Q_t}}{dN_{Q_t}}(y) = \rho_t(x, y), \quad x, y \in H, \ t \geq 0,$$

where

$$\rho_t(x, y) = e^{-\frac{1}{2}|\Gamma(t)x|^2 + \langle \Gamma(t)x, Q_t^{-1/2}y \rangle}, \quad x, y \in H, \ t \geq 0.$$

Consequently, we can write

$$R_t\varphi(x) = \int_H \varphi(y) e^{-\frac{1}{2}|\Gamma(t)x|^2 + \langle \Gamma(t)x, Q_t^{-1/2}y \rangle} N_{Q_t}(dy), \quad x, y \in H, \ t \geq 0.$$

From this identity it follows easily that $R_t\varphi$ is differentiable and that for any $h \in H$ we have

$$\langle D_x R_t \varphi(x), h \rangle = \int_H \varphi(y) \left[\langle \Gamma(t)h, Q_t^{-1/2} y \rangle - \langle \Gamma(t)h, \Gamma(t)x \rangle \right]$$

$$\times \rho_t(x, y) N_{Q_t}(dy),$$

$$= \int_H \langle \Gamma(t)h, Q_t^{-1/2} y \rangle \varphi(e^{tA}x + y) N_{Q_t}(dy). \qquad (8.22)$$

Thus $R_t \varphi \in C_b^1(H)$. Iterating this procedure, we can prove that $R_t \varphi \in C_b^\infty(H)$.

$(ii) \Rightarrow (i)$. Assume that (ii) holds and, by contradiction, that there exists $x_0 \in H$ such that

$$e^{tA}x_0 \notin Q_t^{1/2}(H).$$

Then, again by the Cameron–Martin theorem it follows that, for all $n \in \mathbb{N}$, the measures $N_{e^{tA}x_0/n, Q_t}$ and N_{Q_t} are singular. Thus for any $n \in \mathbb{N}$ there exists a Borel subset K_n of H such that

$$N_{e^{tA}x_0/n, Q_t}(K_n) = 0, \quad N_{Q_t}(K_n) = 1, \quad n \in \mathbb{N}.$$

Setting $K = \bigcap_{n=1}^\infty K_n$, it follows that

$$N_{e^{tA}x_0/n, Q_t}(K) = 0, \quad N_{Q_t}(K) = 1, \quad n \in \mathbb{N}.$$

Consequently we have

$$R_t \mathbf{1}_K(x_0/n) = \int_K N_{e^{tA}x_0/n, Q_t}(dy) = 0, \quad n \in \mathbb{N},$$

whereas

$$R_t \mathbf{1}_K(0) = \int_K N_{Q_t}(dy) = 1, \quad n \in \mathbb{N}.$$

Therefore the function $R_t \mathbf{1}_K$ is not continuous at 0, which contradicts the statement (ii). $\square$

Remark 8.17 Condition (8.18) holds if and only if R_t is strong Feller, see [27].

Example 8.18 Let $H = \mathbb{R}^2$, and

$$A = \begin{pmatrix} 0 & 0 \\ 1 & 0 \end{pmatrix}, \quad C = \begin{pmatrix} 1 & 0 \\ 0 & 0 \end{pmatrix}.$$

Then we have

$$e^{tA} = \begin{pmatrix} 1 & 0 \\ t & 1 \end{pmatrix}, \quad e^{tA}Ce^{tA^*} = \begin{pmatrix} 1 & t \\ t & t^2 \end{pmatrix}.$$

It follows that

$$\int_0^t e^{sA}Ce^{sA^*}\,ds = \int_0^t \begin{pmatrix} 1 & s \\ s & s^2 \end{pmatrix} ds = \begin{pmatrix} t & t^2/2 \\ t^2/2 & t^3/3 \end{pmatrix}.$$

Therefore $\det Q_t > 0$ and so assumption (8.18) is fulfilled. Consequently the solution of the Kolmogorov equation

$$\begin{cases} D_t u(t, x_1, x_2) = \dfrac{1}{2} D_{x_1}^2 u(t, x_1, x_2) + x_1 D_{x_2} u(t, x_1, x_2) := Lu(t, x_1, x_2) \\[2mm] u(0, x) = \varphi(x), \quad \varphi \in B_b(\mathbb{R}^2), \end{cases}$$

(8.23)

is given by $u(t, x) = R_t \varphi(x)$, and by Theorem 8.16 $u(t, \cdot)$ is of class C^∞ for any $t > 0$. This proves that the operator L above is hypoelliptic. One can show indeed that (8.18) reduces, in the present case, to the hypoellipticity condition due to Hörmander.

8.3.2 Invariant measures

We assume here that the linear operator A is of negative type, i.e. there exist $M > 0, \omega > 0$ such that

$$\|e^{tA}\| \le Me^{-\omega t}, \quad t \ge 0. \tag{8.24}$$

Under this assumption the linear operator

$$Q_\infty x = \int_0^{+\infty} e^{tA}Ce^{tA^*}x\,dt, \quad x \in H,$$

is well defined.

Lemma 8.19 Q_∞ is of trace class.

Proof. We have in fact for any $x \in H$,

$$\begin{aligned} Q_\infty x &= \sum_{k=1}^\infty \int_{k-1}^k e^{(s+k-1)A}Ce^{(s+k-1)A^*}x\,ds \\ &= \sum_{k=1}^\infty e^{(k-1)A}Q_1 e^{(k-1)A^*}x\,ds. \end{aligned}$$

It follows that

$$\mathrm{Tr}\, Q_\infty \le M \sum_{k=1}^\infty e^{-2\omega(k-1)}\,\mathrm{Tr}\, Q_1.$$

$\square$

Theorem 8.20 *Assume that Hypothesis 8.10 holds. A probability measure $\mu \in \mathscr{P}(H)$ is invariant for R_t if and only if $\mu = N_{Q_\infty}$. If $\mu = N_{Q_\infty}$ then for any $\varphi \in L^2(H,\mu)$ we have*

$$\lim_{t \to +\infty} R_t\varphi(x) = \int_H \varphi(y)N_{Q_\infty}(dy) \quad \text{in } L^2(H,\mu), \quad x \in H, \qquad (8.25)$$

thus μ is ergodic and strongly mixing.

Proof. *Existence.* We prove that $\mu = N_{Q_\infty}$ is invariant. For this it is enough to check that

$$\int_H R_t\varphi_h(x)\mu(dx) = \int_H \varphi_h(x)\mu(dx), \qquad (8.26)$$

for all $\varphi(x) = e^{i\langle x,h\rangle}$, $h \in H$. In fact (8.26) is equivalent to

$$\langle Q_\infty e^{tA^*}h, e^{tA^*}h\rangle + \langle Q_t h, h\rangle = \langle Q_\infty h, h\rangle, \quad h \in H,$$

which is also equivalent to

$$e^{tA}Q_\infty e^{tA^*} + Q_t = Q_\infty,$$

which can be checked easily.

Uniqueness. Assume that μ is invariant. Then we have

$$\hat{\mu}(e^{tA^*}h)\, e^{-\frac{1}{2}\langle Q_t h,h\rangle} = \hat{\mu}(h), \quad t \geq 0, \ h \in H,$$

where $\hat{\mu}$ is the Fourier tranform of μ. As $t \to \infty$ we find

$$\hat{\mu}(h) = e^{-\frac{1}{2}\langle Q_\infty h,h\rangle}.$$

This implies, by the uniqueness of the Fourier transform of a measure, that $\mu = N_{Q_\infty}$.

Finally, let us show (8.25). It is enough to take an arbitrary exponential function $\varphi = \varphi_h$, $h \in H$. In this case we have by (8.14)

$$\lim_{t \to \infty} R_t\varphi_h(x) = \lim_{t \to \infty} e^{-\frac{1}{2}\langle Q_t h,h\rangle} e^{i\langle e^{tA^*}x,h\rangle} = e^{-\frac{1}{2}\langle Q_\infty h,h\rangle}.$$

Since

$$e^{-\frac{1}{2}\langle Q_\infty h,h\rangle} = \int_H \varphi_h(x)N_{Q_\infty}(dx),$$

the conclusion follows. $\square$

By Theorem 5.8 we obtain the following result.

Proposition 8.21 *For any $p \geq 1$, R_t has a unique extension to a strongly continuous semigroup of contractions in $L^p(H,\mu)$ (which we still denote by R_t).*

For all $p \geq 1$ we shall denote by L_p the infinitesimal generator of R_t in $L^p(H, \mu)$ and by $D(L_p)$ its domain. Let $h \in H$, then, by the very definition of the infinitesimal generator (see (A.4)), we see that [3]

$$\varphi_h \in D(L_p) \iff h \in D(A^*).$$

Moreover if $h \in D(A^*)$ we have

$$L_p \varphi_h(x) = \left[-\frac{1}{2} \langle Ch, h \rangle + i \langle A^* h, x \rangle \right] e^{i \langle h, x \rangle} \varphi_h(x)$$

$$= \frac{1}{2} \operatorname{Tr} [C D_x^2 \varphi_h(x)] + \langle x, A^* D_x \varphi_h(x) \rangle, \quad x \in H. \qquad (8.27)$$

This identity prompts us to introduce the following subspace of $\mathscr{E}(H)$,

$$\mathscr{E}_A(H) := \text{linear span } \{\Re e \, \varphi_h, \ \Im m \, \varphi_h, \ \varphi_h(x) = e^{i \langle h, x \rangle} : \ h \in D(A^*)\}. \qquad (8.28)$$

It is easy to see that $\mathscr{E}_A(H)$ is stable for R_t and it is dense in $L^p(H, \mu)$ for all $p \geq 1$.

Theorem 8.22 *For any $p \geq 1$, $\mathscr{E}_A(H)$ is a core [4] for L_p. Moreover,*

$$L_p \varphi(x) = \frac{1}{2} \operatorname{Tr} [C D_x^2 \varphi(x)] + \langle x, A^* D_x \varphi(x) \rangle, \quad x \in H, \ \varphi \in \mathscr{E}_A(H).$$

Proof. Since $\mathscr{E}_A(H)$ is invariant for R_t and dense in $L^p(H, \mu)$, it follows that it is a core for L_p, by Proposition A.19. The above expression for L_p follows from (8.27). $\square$

---

[3] Recall that $\varphi_h(x) = e^{i \langle h, x \rangle}$, $x \in H$.
[4] See subsection A.3.1

L^2 spaces with respect to a Gaussian measure

9.1 Notations

We are given a separable real Hilbert space H (norm $|\cdot|$, inner product $\langle \cdot, \cdot \rangle$), and a nondegenerate Gaussian measure $\mu = N_Q$ on H. We denote by (e_k) a complete orthonormal system in H and by (λ_k) a sequence of positive numbers such that

$$Qe_k = \lambda_k e_k, \quad k \in \mathbb{N}. \tag{9.1}$$

We set

$$P_n x = \sum_{k=1}^{n} \langle x, e_k \rangle e_k, \quad n \in \mathbb{N} \tag{9.2}$$

and, for any $x \in H$, $x_k = \langle x, e_k \rangle$, $k \in \mathbb{N}$.

We denote by $L^2(H, \mu)$ the Hilbert space of all equivalence classes of Borel square integrable real functions on H, endowed with the inner product,

$$\langle \varphi, \psi \rangle_{L^2(H,\mu)} = \int_H \varphi \, \psi \, d\mu, \quad \varphi, \psi \in L^2(H, \mu).$$

For any $\varphi \in L^2(H, \mu)$ we set

$$\|\varphi\|_{L^2(H,\mu)} = \left(\int_H |\varphi(x)|^2 \mu(dx) \right)^{1/2}.$$

Finally, we denote by $L^2(H, \mu; H)$ the space of all equivalence classes of Borel square integrable mappings $F \colon H \to H$, such that

$$\|F\|_{L^2(H,\mu;H)} := \left(\int_H |F(x)|^2 \mu(dx) \right)^{1/2} < +\infty.$$

The space $L^2(H, \mu; H)$, endowed with the inner product,

$$\langle F, G \rangle_{L^2(H,\mu;H)} = \int_H \langle F(x), G(x) \rangle \mu(dx), \quad F, G \in L^2(H, \mu; H),$$

is a Hilbert space. The elements of $L^2(H, \mu; H)$ are called L^2 *vector fields*.

In this chapter an important rôle will be played by the white noise mapping $W: H \rightarrow L^2(H, \mu)$ introduced in section 1.7. We will use frequently the following identities

$$\int_H W_f W_g \, d\mu = \langle f, g \rangle, \quad f, g \in H, \tag{9.3}$$

$$\int_H e^{W_f} \, d\mu = e^{\frac{1}{2}|f|^2}, \quad f \in H. \tag{9.4}$$

We shall also consider the space $\mathscr{E}(H)$ of exponential functions introduced in section 8.1. By Lemma 8.1 and the dominated convergence theorem it follows that $\mathscr{E}(H)$ is dense in $L^2(H, \mu)$.

Exercise 9.1 Show that the linear span of all real and imaginary parts of functions

$$\varphi_j(x) = e^{i\langle x, e_j \rangle}, \quad j \in \mathbb{N},$$

is dense in $L^2(H, \mu)$.

The content of this chapter is as follows. Section 9.2 is devoted to the construction of a canonical orthonormal basis in $L^2(H, \mu)$ and section 9.3 to the Wiener–Itô decomposition. Finally, in section 9.4 we introduce the classical *Ornstein–Uhlenbeck semigroup* in $L^2(H, \mu)$ and prove the *Mehler formula*.

9.2 Orthonormal basis in $L^2(H, \mu)$

We shall first consider the case when $H = \mathbb{R}$ and $\mu = N_1$. Then we consider the general case.

9.2.1 The one-dimensional case

Here we take $H = \mathbb{R}$ and $\mu = N_1$. An orthonormal basis on $L^2(H, \mu)$ will be defined in terms of the *Hermite polynomials* that we introduce now.

Let us consider the analytic function in the variables $(t, \xi) \in \mathbb{R}^2$ defined as

$$F(t, \xi) = e^{-\frac{t^2}{2} + t\xi}, \quad t, \xi \in \mathbb{R}$$

and let $(H_n)_{n \in \{0\} \cup \mathbb{N}}$ be such that

$$F(t, \xi) = \sum_{n=0}^{\infty} \frac{t^n}{\sqrt{n!}} H_n(\xi), \quad t, \xi \in \mathbb{R}. \tag{9.5}$$

Proposition 9.2 *For any* $n \in \{0\} \cup \mathbb{N}$ *the following identity holds*

$$H_n(\xi) = \frac{(-1)^n}{\sqrt{n!}} e^{\frac{\xi^2}{2}} D_\xi^n \left(e^{-\frac{\xi^2}{2}}\right), \quad \xi \in \mathbb{R}. \tag{9.6}$$

Proof. Write

$$F(t, \xi) = e^{\frac{\xi^2}{2}} e^{-\frac{1}{2}(t - \xi)^2} = \sum_{n=0}^{\infty} \frac{t^n}{n!} e^{\frac{\xi^2}{2}} D_t^n \left(e^{-\frac{1}{2}(t-\xi)^2}\right)\Big|_{t=0}$$

$$= \sum_{n=0}^{\infty} \frac{t^n}{n!} (-1)^n e^{\frac{\xi^2}{2}} D_\xi^n \left(e^{-\frac{\xi^2}{2}}\right).$$

Now the conclusion follows comparing this identity with (9.5). $\square$

By Proposition 9.2 we see that for any $n \in \mathbb{N} \cup \{0\}$, H_n is a polynomial of degree n with a positive leading coefficient. H_n are called *Hermite* polynomials. We have in particular

$$H_0(\xi) = 1, \quad H_1(\xi) = \xi, \quad H_2(\xi) = \frac{1}{\sqrt{2}} (\xi^2 - 1),$$

$$H_3(\xi) = \frac{1}{\sqrt{6}} (\xi^3 - 3\xi), \quad H_4(\xi) = \frac{1}{2\sqrt{6}} (\xi^4 - 6\xi^2 + 3).$$

In the following proposition are collected some important properties of the Hermite polynomials.

Proposition 9.3 *For any* $n \in \mathbb{N}$ *we have*

$$\xi H_n(\xi) = \sqrt{n+1}\, H_{n+1}(\xi) + \sqrt{n}\, H_{n-1}(\xi), \quad \xi \in \mathbb{R}, \tag{9.7}$$

$$D_\xi H_n(\xi) = \sqrt{n}\, H_{n-1}(\xi), \quad \xi \in \mathbb{R}, \tag{9.8}$$

$$D_\xi^2 H_n(\xi) - \xi D_\xi H_n(\xi) = -n H_n(\xi), \quad \xi \in \mathbb{R}. \tag{9.9}$$

Proof. Equations (9.7) and (9.8) follow from the identities

$$D_t F(t, \xi) = (\xi - t) F(t, \xi) = \sum_{n=1}^{\infty} \sqrt{n} \, \frac{t^{n-1}}{\sqrt{(n-1)!}} \, H_n(\xi), \quad t, \xi \in \mathbb{R},$$

and

$$D_\xi F(t, \xi) = t F(t, \xi) = \sum_{n=1}^{\infty} \frac{t^n}{\sqrt{n!}} \, H_n'(\xi), \quad t, \xi \in \mathbb{R},$$

respectively. Finally, equation (9.9) is an immediate consequence of (9.7) and (9.8). □

Identity (9.8) shows that the derivation D_ξ acts as a shift operator with respect to the system $(H_n)_{n \in \{0\} \cup \mathbb{N}}$. Moreover by (9.9) it follows that the Hermite operator

$$\mathcal{H} \, \varphi := \frac{1}{2} \, D_\xi^2 \varphi - \frac{1}{2} \, \xi D_\xi \varphi,$$

is diagonal with respect to system $(H_n)_{n \in \{0\} \cup \mathbb{N}}$.

Proposition 9.4 *System $(H_n)_{n \in \{0\} \cup \mathbb{N}}$ is orthonormal and complete in $L^2(\mathbb{R}, \mu)$, where $\mu = N_1$.*

Proof. We first prove orthonormality. Write

$$F(t, \xi) F(s, \xi) = e^{-\frac{1}{2}(t^2 + s^2) + \xi(t+s)}$$

$$= \sum_{m,n=0}^{\infty} \frac{t^m}{\sqrt{m!}} \frac{s^n}{\sqrt{n!}} \, H_n(\xi) \, H_m(\xi), \quad t, s, \xi \in \mathbb{R}.$$

Integrating this identity with respect to μ and taking into account that

$$\int_{\mathbb{R}} F(t, \xi) F(s, \xi) \mu(d\xi) = (2\pi)^{-1/2} e^{-\frac{1}{2}(t^2 + s^2)} \int_{\mathbb{R}} e^{-\frac{\xi^2}{2}} e^{\xi(t+s)} d\xi = e^{ts},$$

we find that

$$e^{ts} = \sum_{m,n=0}^{\infty} \frac{t^m}{\sqrt{m!}} \frac{s^n}{\sqrt{n!}} \int_{\mathbb{R}} H_n(\xi) \, H_m(\xi) \, \mu(d\xi), \quad t, s \in \mathbb{R},$$

which yields

$$\int_{\mathbb{R}} H_n(\xi) \, H_m(\xi) \, \mu(d\xi) = \delta_{n,m}, \quad n, m \in \{0\} \cup \mathbb{N}.$$

It remains to prove the completeness of (H_n). Let $f \in L^2(\mathbb{R}, \mu)$ be such that

$$\int_{\mathbb{R}} f(\xi) H_n(\xi) \mu(d\xi) = 0 \quad \text{for all } n \in \{0\} \cup \mathbb{N}.$$

Then, noticing that $g(\xi) := f(\xi)e^{-\frac{\xi^2}{4}}$ belongs to $L^2(\mathbb{R})$, we have by (9.5)

$$0 = \int_{\mathbb{R}} f(\xi)F(t,\xi)\mu(d\xi) = (2\pi)^{-1/2}e^{\frac{t^2}{2}}\int_{\mathbb{R}} g(\xi)e^{-\frac{(2t-\xi)^2}{4}}\,d\xi.$$

which implies $g * \varphi_0 = 0$, where $*$ means convolution and

$$\varphi_0(\xi) = e^{-\frac{\xi^2}{4}}, \quad \xi \in \mathbb{R}.$$

Taking the Fourier transform of both sides of the identity $g * \varphi_0 = 0$, yields

$$\hat{g}(\eta)\hat{\varphi}_0(\eta) = \hat{g}(\eta)e^{-\frac{\eta^2}{4}} = 0, \quad \eta \in \mathbb{R}$$

and so, $f = 0$. $\square$

9.2.2 The infinite dimensional case

We shall construct here a complete orthonormal system on $L^2(H, \mu)$ in terms of generalized Hermite polynomials. We shall need the following basic lemma

Lemma 9.5 *Let $h, g \in H$ with $|h| = |g| = 1$ and let $n, m \in \mathbb{N} \cup \{0\}$. Then we have*

$$\int_H H_n(W_h)H_m(W_g)d\mu = \delta_{n,m}[\langle h, g\rangle]^n. \tag{9.10}$$

Proof. For any $t, s \in \mathbb{R}$ we have

$$\int_H F(t, W_h)F(s, W_g)d\mu = e^{-\frac{t^2+s^2}{2}}\int_H e^{tW_h+sW_g}\,d\mu$$

$$= e^{-\frac{t^2+s^2}{2}}\int_H e^{W_{th+sg}}\,d\mu = e^{-\frac{t^2+s^2}{2}}e^{\frac{1}{2}|th+sg|^2} = e^{ts\langle h,g\rangle},$$

because $|h| = |g| = 1$. It follows that

$$e^{ts\langle h,g\rangle} = \sum_{m,n=0}^{\infty}\frac{t^n s^m}{\sqrt{n!m!}}\int_H H_n(W_h)H_m(W_g)\,d\mu.$$

Since

$$e^{ts\langle h,g\rangle} = \sum_{m=0}^{\infty}\frac{\langle h, g\rangle^m}{m!}t^m s^m,$$

(9.10) follows. $\square$

Exercise 9.6 Prove that for any $m \in \mathbb{N}$ there exists $C_m > 0$ such that

$$\int_H |W_f|^{2m} d\mu \le C_m(1 + |f|^m). \tag{9.11}$$

Hint: Express the polynomial ξ^{2m} as a linear combination of Hermite polynomials and use (9.10).

We are now ready to define a complete orthonormal system in $L^2(H, \mu)$. Let Γ be the set of all mappings

$$\gamma : \mathbb{N} \to \{0\} \cup \mathbb{N}, \; n \to \gamma_n,$$

such that

$$|\gamma| := \sum_{k=1}^{\infty} \gamma_k < +\infty.$$

Note that if $\gamma \in \Gamma$ then $\gamma_n = 0$ for all n, except possibly a finite number. For any $\gamma \in \Gamma$ we define the *Hermite polynomial*,

$$H_\gamma(x) = \prod_{k=1}^{\infty} H_{\gamma_k}(W_{e_k}(x)), \quad x \in H.$$

This definition is meaningful since each factor, with the exception of at most a finite number, is equal to $H_0(W_{e_k}(x)) = 1$, $x \in H$.

We can now prove the result.

Theorem 9.7 *System* $(H_\gamma)_{\gamma \in \Gamma}$ *is orthonormal and complete on* $L^2(H, \mu)$.

Proof. *Orthonormality.* Let $\gamma, \eta \in \Gamma$. Then there exists $N \in \mathbb{N}$ such that

$$H_\gamma(x) = \prod_{k=1}^{N} H_{\gamma_k}(W_{e_k}(x)), \quad H_\eta(x) = \prod_{k=1}^{N} H_{\eta_k}(W_{e_k}(x)).$$

Consequently,

$$\int_H H_\gamma H_\eta d\mu = \int_H \prod_{n=1}^{N} H_{\gamma_n}(W_{e_n}) H_{\eta_n}(W_{e_n}) d\mu$$

$$= \prod_{n=1}^{N} \int_H H_{\gamma_n}(W_{e_n}) H_{\eta_n}(W_{e_n}) d\mu = \delta_{\gamma\eta}$$

(where $\delta_{\eta,\gamma} = \prod_{n=1}^{N} \delta_{\eta_n,\gamma_n}$), since $W_{e_1}, \ldots, W_{e_N}$ are independent by Proposition 1.28.

Completeness. Let $\varphi_0 \in L^2(H, \mu)$ be such that

$$\int_H \varphi_0 H_\gamma d\mu = 0 \quad \text{for all } \gamma \in \Gamma. \tag{9.12}$$

We have to show that $\varphi_0 = 0$.

Let us consider the signed measure on $(H, \mathscr{B}(H))$ defined by $\zeta(dx) := \psi_0(x)\mu(dx)$ and let $\zeta_n = (P_n)_\# \zeta$. It is enough to show that $\zeta_n = 0$ for all $n \in \mathbb{N}$. We shall first prove that

$$F(h) := \int_{P_n(H)} e^{\langle P_n h, \xi \rangle} \zeta_n(d\xi)$$

$$= \int_H e^{\langle P_n h, x \rangle} \psi_0(x)\mu(dx) = 0, \quad \forall\, h \in P_n(H). \tag{9.13}$$

Let $h \in P_n(H)$. By (9.12) it follows that

$$\int_H F(t_1, W_{e_1}) \dots F(t_n, W_{e_n}) \varphi_0 d\mu = 0, \quad \forall\, t_1, \dots, t_n \in \mathbb{R}^n,$$

which implies that,

$$\int_H e^{t_1 W_{e_1} + \dots + t_n W_{e_n}} \varphi_0 d\mu = \int_H e^{t_1 \lambda_1^{-1/2} x_1 + \dots + t_1 \lambda_n^{-1/2} x_n} \varphi_0 d\mu = 0,$$

so that (9.13) follows from the arbitrariness of $t_1, \dots, t_n$.

To conclude the proof we notice that the function

$$F(\xi_1, \dots, \xi_n) = \int_H e^{\xi_1 x_1 + \dots + \xi_n x_n} \psi_0(x)\mu(dx), \quad (\xi_1, \dots, \xi_n) \in \mathbb{R}^n,$$

is analytic in $\mathbb{R}^n$, since for all $(k_1, \dots, k_n) \in \mathbb{N}^n$ we have, setting $k = k_1 + \dots + k_n$,

$$\frac{\partial^k}{\partial_{h_1}^{k_1} \dots \partial_{h_n}^{k_n}} F(h_1, \dots, h_n) = \int_H x_1^{k_1} \dots x_n^{k_n} e^{h_1 x_1 + \dots + h_n x_n} \psi_0(x)\mu(dx).$$

(Note that the integral above is convergent since the function $x \to x_1^{k_1} \dots x_n^{k_n}$ belongs to $L^2(H, \mu)$.)

Since F is analytic by (9.13) it follows that the Fourier transform $F(ih)$ of ζ_n vanishes so that ζ_n vanishes as well. In conclusion we have proved that $\zeta_n = 0$ for all $n \in \mathbb{N}$ so that $\psi_0 = 0$ in $L^2(H, \mu)$, as required. $\square$

9.3 Wiener–Itô decomposition

For all $n \in \{0\} \cup \mathbb{N}$ we denote by $L_n^2(H, \mu)$ the closed subspace of $L^2(H, \mu)$ spanned by

$$\{H_n(W_f) : f \in H, |f| = 1\}.$$

In particular $L_0^2(H, \mu)$ is the space of all constants and $L_1^2(H, \mu)$ is the space of all Gaussian random variables which belong to H.

Exercise 9.8 Prove that

$$L_1^2(H, \mu) = \mathrm{span}\ \{W_f : f \in H\}.$$

We shall denote by Π_n the orthogonal projector of $L^2(H, \mu)$ on $L_n^2(H, \mu)$, $n \in \{0\} \cup \mathbb{N}$. Arguing as in the proof of Theorem 9.7 we see that

$$L^2(H, \mu) = \bigoplus_{n=0}^{\infty} L_n^2(H, \mu). \tag{9.14}$$

Formula (9.14) is called the *Wiener–Itô decomposition* or the *cahos decomposition* of $L^2(H, \mu)$ and $L_n^2(H, \mu)$ is called the nth component of the decomposition. Notice that $L_n^2(H, \mu)$ does not depend on the choice of the basis (e_k).

Let us give an interesting characterization of $L_n^2(H, \mu)$, $n \in \{0\} \cup \mathbb{N}$.

Proposition 9.9 *For any $n \in \{0\} \cup \mathbb{N}$ the space $L_n^2(H, \mu)$ coincides with the closed subspace of $L^2(H, \mu)$ spanned by*

$$V_n := \{H_\gamma : |\gamma| = n\}.$$

Proof. It is enough to show that if $n, N \in \mathbb{N}$, $f \in H$ with $|f| = 1$, $k_1, \ldots, k_N \in \mathbb{N}$, and $k_1 + \cdots + k_N \neq n$, we have

$$\int_H H_{k_1}(W_{e_1})...H_{k_N}(W_{e_N})H_n(W_f)d\mu = 0. \tag{9.15}$$

We have in fact

$$I := \int_H F(t_1, W_{e_1})...F(t_N, W_{e_N})F(t_{N+1}, W_f)d\mu$$

$$= e^{-\frac{1}{2}(t_1^2 + \cdots + t_{N+1}^2)} \int_H e^{W_{t_1 e_1 + \cdots + t_N e_N + t_{N+1}f}}d\mu$$

$$= e^{t_{N+1}(t_1 f_1 + \cdots + t_N f_N)}.$$

On the other hand, since

$$I = \sum_{k_1, \ldots, k_{N+1}=0}^{\infty} \frac{t_1^{k_1} \ldots t_{N+1}^{k_{N+1}}}{\sqrt{k_1! \ldots k_{N+1}!}} \int_H H_{k_1}(W_{e_1})...H_{k_N}(W_{e_N})H_{k_{N+1}}(W_f)d\mu,$$

the conclusion follows. $\square$

Remark 9.10 Let $p(\xi)$ be a real polynomial of degree $n \in \mathbb{N}$, and let $f \in H$ with $|f| = 1$. Then we have

$$p(W_f) \in \bigoplus_{k=0}^{n} L_n^2(H, \mu). \tag{9.16}$$

In fact there exists $c_1, \ldots, c_n \in \mathbb{R}$ such that $p(\xi) = \sum_{k=0}^{n} c_k H_k(\xi)$, and so $p(W_f) = \sum_{k=0}^{n} c_k H_k(W_f)$, which implies (9.16).

We now prove an important property of the projection Π_n.

Proposition 9.11 Let $f \in H$ such that $|f| = 1$, and let $n, k \in \mathbb{N}$. Then we have

$$\Pi_k(W_f^n) = \begin{cases} \sqrt{n!} \begin{pmatrix} k \\ n \end{pmatrix} H_n(W_f), & \text{if } k \geq n, \\[2mm] 0, & \text{if } k < n. \end{cases} \tag{9.17}$$

Proof. We give the proof for $k = n$; the other cases can be handled in a similar way. Since $\sqrt{n!}\, H_n(W_f) \in L_n^2(H, \mu)$ by definition, it is enough to show that for all $g \in H$ such that $|g| = 1$, we have

$$\int_H [W_f^n - \sqrt{n!}\ H_n(W_f)] H_n(W_g) d\mu = 0,$$

or, equivalently, that

$$\int_H W_f^n H_n(W_g) d\mu = \sqrt{n!}\ [\langle f, g \rangle]^n. \tag{9.18}$$

Now (9.18) follows easily from the identity

$$I := \int_H e^{sW_f} H_n(W_g) d\mu = \frac{1}{\sqrt{n!}}\ s^n e^{\frac{s^2}{2}}\ [\langle f, g \rangle]^n \tag{9.19}$$

(by differentiating n times with respect to s and then setting $s = 0$), which we shall prove now. We have, taking into account (9.10),

$$I = e^{\frac{s^2}{2}} \int_H F(s, W_f) H_n(W_g) d\mu$$

$$= e^{\frac{s^2}{2}} \sum_{k=0}^{\infty} \frac{s^k}{\sqrt{k!}} \int_H H_k(W_f) H_n(W_g) d\mu$$

$$= \frac{1}{\sqrt{n!}}\ s^n e^{\frac{s^2}{2}}\ [\langle f, g \rangle^n],$$

which yields (9.19). $\square$

Now we can compute easily the projections of an exponential function on $L_n^2(H, \mu)$.

Corollary 9.12 *Let $f \in H$ with $|f| = 1$. Then we have*

$$\Pi_n \left(e^{sW_f} \right) = \frac{1}{\sqrt{n!}} \, s^n e^{\frac{s^2}{2}} \, H_n(W_f), \quad n \in \mathbb{N}. \tag{9.20}$$

Proof. We have in fact

$$\Pi_n \left(e^{sW_f} \right) = \sum_{k=0}^{\infty} \frac{s^k}{k!} \, \Pi_n(W_f^k) = s^n e^{\frac{s^2}{2}} \, H_n(W_f).$$

$\square$

9.4 The classical Ornstein–Uhlenbeck semigroup

We define a semigroup of linear bounded operators on $L^2(H; \mu)$ by setting

$$U_t \varphi = \sum_{n=0}^{\infty} e^{-\frac{nt}{2}} \, \Pi_n \varphi, \quad t \geq 0, \ \varphi \in L^2(H; \mu), \tag{9.21}$$

where Π_n is the orthogonal projection on the nth component of the Wiener–Itô decomposition of $L^2(H, \mu)$.

It is easy to see that U_t is a strongly continuous semigroup, called the *classical Ornstein–Uhlenbeck semigroup*. Its infinitesimal generator L is given by

$$\begin{cases} D(L) = \left\{ \varphi \in L^2(H, \mu) : \ \sum_{n=0}^{\infty} n^2 |\Pi_n \varphi|^2 < +\infty \right\} \\ \\ L\varphi = -\sum_{n=0}^{\infty} n \, \Pi_n \varphi, \quad \varphi \in D(L). \end{cases} \tag{9.22}$$

The following result shows that the semigroup U_t coincides with the Ornstein–Uhlenbeck semigroup defined by (8.13) with $A = \frac{1}{2} I$ and $C = Q$, introduced in Chapter 8.

Theorem 9.13 *For any $\varphi \in L^2(H, \mu)$ we have*

$$U_t \varphi(x) = \int_H \varphi(e^{-\frac{t}{2}} x + y) N_{(1-e^{-t})Q}(dy). \tag{9.23}$$

Proof. It is enough to check (9.23) for $\varphi = e^{sW_f}$ where $f \in H$ and $|f| = 1$. In this case by (9.20) we have

$$U_t\left(e^{sW_f}\right) = \sum_{n=0}^{\infty} e^{-\frac{nt}{2}} \Pi_n\left(e^{sW_f}\right) = \sum_{n=0}^{\infty} e^{-\frac{nt}{2}} \frac{1}{\sqrt{n!}} s^n e^{\frac{s^2}{2}} H_n(W_f)$$

$$= e^{\frac{s^2}{2}} \sum_{n=0}^{\infty} \frac{1}{\sqrt{n!}} [se^{-\frac{t}{2}}]^n H_n(W_f)$$

$$= e^{\frac{s^2}{2}} F(se^{-t/2}, W_f) = e^{\frac{s^2}{2}(1-e^{-t})} e^{se^{-t/2}W_f}. \tag{9.24}$$

On the other hand, we have

$$\int_H e^{sW_f(e^{-\frac{t}{2}}x+y)} N_{(1-e^{-t})Q}(dy) = e^{se^{-\frac{t}{2}}W_f(x)} \int_H e^{sW_f(y)} N_{(1-e^{-t})Q}(dy)$$

$$= e^{se^{-\frac{t}{2}}W_f(x)} e^{\frac{s^2}{2}(1-e^{-t})},$$

which coincides with (9.24). $\square$

Remark 9.14 Assume that $\varphi \in C_b(H)$. Then setting $y = \sqrt{1-e^{-t}}\, z$, we deduce by (9.23) the *Mehler* formula

$$U_t\varphi(x) = \int_H \varphi(e^{-\frac{t}{2}}x + \sqrt{1-e^{-t}}\, z) N_Q(dz). \tag{9.25}$$

10

Sobolev spaces for a Gaussian measure

We are given a nondegenerate Gaussian measure $\mu = N_Q$ on a separable Hilbert space H. We use here notations of Chapter 9. In particular, we denote by (e_k) a complete orthonormal system in H and by (λ_k) a sequence of positive numbers such that (9.1) holds.

In section 10.1 we shall define the generalized gradient $\overline{D}$ on the Sobolev space $W^{1,2}(H,\mu)$. To define $\overline{D}$ we proceed as follows. First we consider the usual gradient operator defined on the space $\mathscr{E}(H)$ of exponential functions and show that it is closable. Then we define the space $W^{1,2}(H,\mu)$ as the domain of the closure of the gradient. [1]

This procedure is the natural generalization of the definition of the gradient (in the sense of Friedrichs) in $L^2(\mathbb{R}^n)$ with respect to the Lebesgue measure, the only difference being that the space $\mathscr{E}(H)$ is replaced there by $C_0^\infty(\mathbb{R}^n)$.

Then we study some properties of the Sobolev space $W^{1,2}(H,\mu)$, we present a characterization of $W^{1,2}(H,\mu)$ in terms of Wiener–Itô decomposition and prove compactness of the embedding

$$W^{1,2}(H,\mu) \subset L^2(H,\mu).$$

We notice that the analogue of this last result fails in $L^2(\mathbb{R}^n)$.

In section 10.2 we shall express the space $W^{1,2}(H,\mu)$ in terms of the Wiener chaos and in section 10.3 we shall study the adjoint D^* of $\overline{D}$. Section 10.4 is devoted to the Dirichlet form associated to μ and to the corresponding semigroup R_t. It happens that R_t belongs to the class of the Ornstein–Uhlenbeck semigroups introduced in section 8.3. With the help of R_t we prove in section 10.5, following [13], the Poincaré and

---

[1] For concepts of closure and closability of linear operators see Appendix A.

log–Sobolev inequalities corresponding to the measure μ. As a consequence, we obtain the hypercontractivity of R_t. Finally, in section 10.6 we introduce the Sobolev space $W^{2,2}(H,\mu)$ and characterize the domain of the infinitesimal generator of R_t.

We shall only consider spaces $W^{k,2}(H,\mu)$, $k = 1,2$ for brevity. All results can be generalized, with obvious modifications, to cover Sobolev spaces $W^{k,p}(H,\mu)$, $k = 1,2$, $p \geq 1$.

Throughout this chapter we shall write $D_x = D$ for brevity.

10.1 Derivatives in the sense of Friedrichs

For any $\varphi \in \mathscr{E}(H)^{(2)}$ and any $k \in \mathbb{N}$ we denote by $D_k\varphi$ the derivative of φ in the direction of e_k, namely

$$D_k\varphi(x) = \lim_{\varepsilon \to 0} \frac{1}{\varepsilon}[\varphi(x+\varepsilon e_k) - \varphi(x)], \quad x \in H.$$

Our first goal is to show that the linear mapping

$$D: \mathscr{E}(H) \subset L^2(H,\mu) \to L^2(H,\mu;H), \quad \varphi \mapsto D\varphi,$$

is closable. For this we need an integration by parts formula.

Lemma 10.1 *Let $\varphi, \psi \in \mathscr{E}(H)$. Then the following identity holds,*

$$\int_H D_k\varphi\,\psi\,d\mu = -\int_H \varphi\,D_k\psi\,d\mu + \frac{1}{\lambda_k}\int_H x_k\varphi\,\psi\,d\mu. \tag{10.1}$$

Proof. In view of Lemma 8.1 and the dominated convergence theorem, it is enough to prove (10.1) for

$$\varphi(x) = e^{i\langle f,x\rangle}, \quad \psi(x) = e^{i\langle g,x\rangle}, \quad x \in H,$$

where f and g are arbitrary elements of H. In this case we have by (9.4)

$$\int_H D_k\varphi\,\psi\,d\mu = if_k e^{-\frac{1}{2}\langle Q(f+g),f+g\rangle}, \tag{10.2}$$

$$\int_H \varphi D_k\psi\,d\mu = ig_k e^{-\frac{1}{2}\langle Q(f+g),f+g\rangle}. \tag{10.3}$$

---

(2) See section 8.1

Moreover

$$\int_H x_k \varphi\psi d\mu = \int_H x_k e^{i\langle f+g,x\rangle} \mu(dx) = -i\frac{d}{dt}\int_H e^{i\langle f+g+te_k,x\rangle} \mu(dx)\Big|_{t=0}$$

$$= -i\frac{d}{dt} \, e^{-\frac{1}{2}\langle Q(f+g+te_k),f+g+te_k\rangle}\Big|_{t=0}$$

$$= i\langle Q(f+g),e_k\rangle e^{-\frac{1}{2}\langle Q(f+g),f+g\rangle}$$

$$= i\lambda_k(f_k+g_k)e^{-\frac{1}{2}\langle Q(f+g),f+g\rangle}. \tag{10.4}$$

Now summing up (10.2) and (10.3), yields (10.4). $\square$

Corollary 10.2 *Let* $\varphi,\psi \in \mathscr{E}(H)$ *and* $z \in Q^{1/2}(H)$. *Then the following identity holds,* [3]

$$\int_H \langle D\varphi,z\rangle \psi \, d\mu = -\int_H \langle D\psi,z\rangle \varphi \, d\mu + \int_H W_{Q^{-1/2}z} \, \varphi\,\psi \, d\mu. \tag{10.5}$$

Proof. Let $n \in \mathbb{N}$. Then by (10.1) we have

$$\int_H \langle D\varphi,P_n z\rangle \psi \, d\mu = \sum_{k=1}^n z_k \int_H D_k\varphi \, \psi \, d\mu$$

$$= -\sum_{k=1}^n z_k \int_H D_k\psi \, \varphi \, d\mu + \sum_{k=1}^n \frac{z_k}{\lambda_k} \int_H x_k \, \varphi \, \psi d\mu$$

$$= -\int_H \langle D\psi,P_n z\rangle \varphi \, d\mu - \int_H W_{Q^{-1/2}P_n z} \, \varphi \, \psi \, d\mu.$$

Now the conclusion follows, letting $n \to \infty$. $\square$

Proposition 10.3 *The mapping*

$$D\colon \mathscr{E}(H) \subset L^2(H,\mu) \to L^2(H,\mu;H), \quad \varphi \mapsto D\varphi,$$

is closable.

Proof. Let $(\varphi_n) \subset \mathscr{E}(H)$ be such that

$$\varphi_n \to 0 \quad \text{in } L^2(H,\mu), \quad D\varphi_n \to F \quad \text{in } L^2(H,\mu;H),$$

as $n \to \infty$. We have to show that $F = 0$.

[3] The white noise mapping W was defined in section 1.7.

Let $\psi \in \mathscr{E}(H)$ and $z \in Q^{1/2}(H)$. Then by (10.5) we have

$$\int_H \langle D\varphi_n, z \rangle \psi d\mu = -\int_H \langle D\psi, z \rangle \varphi_n d\mu + \int_H W_{Q^{-1/2}z} \varphi_n \psi d\mu. \quad (10.6)$$

Letting $n \to \infty$ we find that

$$\int_H \langle F(x), z \rangle \psi(x) \mu(dx) = 0.$$

This implies that $F = 0$ in view of the arbitrariness of ψ and z. $\square$

We shall denote by $\overline{D}$ the closure of D and by $W^{1,2}(H, \mu)$ its domain of definition. $W^{1,2}(H, \mu)$, endowed with the scalar product

$$\langle \varphi, \psi \rangle_{W^{1,2}(H,\mu)} = \int_H [\varphi\psi + \langle \overline{D}\varphi, \overline{D}\psi \rangle] d\mu,$$

is a Hilbert space.

When no confusion may arise we shall write D instead of $\overline{D}$.

Exercise 10.4 Prove that for any $k \in \mathbb{N}$ the linear operator D_k, defined in $\mathscr{E}(H)$, is closable. Denote by $\overline{D_k}$ its closure. Prove that

$$\langle \overline{D}\varphi, e_k \rangle = \overline{D_k}\varphi, \quad \varphi \in W^{1,2}(H, \mu) \quad (10.7)$$

and that

$$|\overline{D}\varphi(x)|^2 = \sum_{k=1}^{\infty} |D_{k,\mu}\varphi(x)|^2, \quad \mu\text{-a.e.} \quad (10.8)$$

10.1.1 Some properties of $W^{1,2}(H, \mu)$

We are going to prove some useful properties of the space $W^{1,2}(H, \mu)$. First we show that any function of class C^1 which has a suitable growth (together with its derivative), belongs to $W^{1,2}(H, \mu)$ and that its gradient coincides with $\overline{D}\varphi$, μ-almost everywhere. To prove this fact we shall use Lemma 8.1 about the pointwise approximation of functions of $C_b(H)$ by exponential functions and the following straightforward generalization of that lemma, whose proof is left to the reader.

Lemma 10.5 For all $\varphi \in C_b^1(H)$ there exists a two-index sequence $(\varphi_{k,n}) \subset \mathscr{E}(H)$ such that

(i) $\displaystyle \lim_{k \to \infty} \lim_{n \to \infty} \varphi_{k,n}(x) = \varphi(x)$ for all $x \in H$,

(ii) $\displaystyle \lim_{k \to \infty} \lim_{n \to \infty} D\varphi_{k,n}(x) = D\varphi(x)$ for all $x \in H$,

(iii) $\|\varphi_{k,n}\|_0 + \|D\varphi_{k,n}\|_0 \leq \|\varphi\|_0 + \|D\varphi\|_0 + \frac{1}{n}$ for all $n, k \in \mathbb{N}$.

$$(10.9)$$

Now we can show that any function of class C^1 of suitable growth belongs to $W^{1,2}(H, \mu)$.

Proposition 10.6 *Let* $\varphi \in C_b^1(H)$. *Then* $\varphi \in W^{1,2}(H, \mu)$ *and* $\overline{D}\varphi(x) = D\varphi(x)$ *for* μ-*almost all* $x \in H$.

Proof. By Lemma 10.5 there exists a two-index sequence $(\varphi_{n,k}) \subset \mathscr{E}(H)$ such that (10.9) holds. By the dominated convergence theorem we see that

$$\lim_{k \to \infty} \lim_{n \to \infty} \varphi_{k,n} = \varphi \quad \text{in } L^2(H, \mu)$$

and

$$\lim_{k \to \infty} \lim_{n \to \infty} D\varphi_{k,n} = D\varphi \quad \text{in } L^2(H, \mu; H).$$

So, the conclusion follows. $\square$

Exercise 10.7 Let $\varphi \colon H \to \mathbb{R}$ be continuously differentiable. Assume that there exists $\kappa > 0$ and $\varepsilon \in (0, (1/2)\inf_{k \in \mathbb{N}} \lambda_k^{-1})$ such that

$$|\varphi(x)| + |D\varphi(x)| \le \kappa e^{\varepsilon|x|^2}, \quad x \in H.$$

Prove that $\varphi \in W^{1,2}(H, \mu)$ and $\overline{D}\varphi(x) = D\varphi(x)$ for μ-almost all $x \in H$. **Hint:** Recall Proposition 1.13. Then set

$$\varphi_n(x) = \frac{\varphi(x)}{1 + \frac{1}{n} e^{\varepsilon|x|^2}}, \quad x \in H$$

and prove that

$$\varphi_n \to \varphi \quad \text{in } L^2(H, \mu), \quad D\varphi_n \to D\varphi \quad \text{in } L^2(H, \mu; H).$$

10.1.2 Chain rule

Proposition 10.8 *Let* $\varphi \in W^{1,2}(H, \mu)$ *and* $g \in C_b^1(\mathbb{R})$. *Then* $g(\varphi) \in W^{1,2}(H, \mu)$ *and*

$$\overline{D}g(\varphi) = g'(\varphi)\overline{D}\varphi. \tag{10.10}$$

Proof. Let $(\varphi_n) \subset \mathscr{E}(H)$ be such that

$$\lim_{n \to \infty} \varphi_n = \varphi \quad \text{in } L^2(H, \mu), \quad \lim_{n \to \infty} D\varphi_n = \overline{D}\varphi \quad \text{in } L^2(H, \mu; H).$$

Then $g(\varphi_n)$ belongs to $W^{1,2}(H, \mu)$ by Proposition 10.6. Moreover, since $Dg(\varphi_n) = g'(\varphi_n)D\varphi_n$, we have $Dg(\varphi_n) \in L^2(H, \mu; H)$, and so,

$$\lim_{n \to \infty} Dg(\varphi_n) = g'(\varphi)\overline{D}\varphi \quad \text{in } L^2(H, \mu; H).$$

So, (10.10) holds. $\square$

10.1.3 Gradient of a product

Proposition 10.9 *Let $\varphi, \psi \in W^{1,2}(H, \mu)$ and suppose that ψ and $\overline{D}\psi$ are bounded. Then $\varphi\psi \in W^{1,2}(H, \mu)$ and we have*

$$\overline{D}(\varphi\psi) = \overline{D}(\varphi)\psi + \varphi\overline{D}(\psi). \qquad (10.11)$$

Proof. Step 1. We prove the result under the additional assumption that $\varphi \in \mathscr{E}(H)$.

Since $\psi \in W^{1,2}(H, \mu)$ there exists $(\psi_n) \subset \mathscr{E}(H)$ such that,

$$\lim_{n\to\infty} \psi_n = \psi \quad \text{in } L^2(H, \mu), \quad \lim_{n\to\infty} D\psi_n = \overline{D}\psi \quad \text{in } L^2(H, \mu; H).$$

Then we have $D(\varphi\psi_n) = D(\varphi)\psi_n + \varphi D(\psi_n)$ and so,

$$\lim_{n\to\infty} D(\varphi\psi_n) = D(\varphi)\psi + \varphi\overline{D}(\psi) \quad \text{in } L^2(H, \mu; H).$$

This shows that $\varphi\psi \in W^{1,2}(H, \mu)$ and (10.11) holds in this case.

Step 2. We consider the general case $\varphi \in W^{1,2}(H, \mu)$.

Let $(\varphi_n) \subset \mathscr{E}(H)$ such that

$$\lim_{n\to\infty} \varphi_n = \varphi \quad \text{in } L^2(H, \mu), \quad \lim_{n\to\infty} D\varphi_n = \overline{D}\varphi \quad \text{in } L^2(H, \mu; H).$$

By Step 1 we have

$$\overline{D}(\varphi_n\psi) = D\varphi_n\,\psi + \varphi_n\overline{D}\psi.$$

Since ψ and $\overline{D}\psi$ are bounded, it follows that

$$\lim_{n\to\infty} \overline{D}(\varphi_n\psi) = \overline{D}\varphi\,\psi + \varphi\overline{D}\psi \quad \text{in } L^2(H, \mu; H).$$

So, $\varphi\psi \in W^{1,2}(H, \mu)$ and (10.11) is fulfilled. $\square$

10.1.4 Lipschitz continuous functions

We denote by $\mathrm{Lip}(H)$ the set all functions $\varphi \to \mathbb{R}$ such that

$$[\varphi]_1 := \sup_{x,y\in H} \frac{|\varphi(x) - \varphi(y)|}{|x - y|} < +\infty.$$

In this subsection we shall prove that any real Lipschitz continuous function on H belongs to $W^{1,2}(H, \mu)$. To this purpose we recall an analytic result.

Proposition 10.10 *Assume that E is a Hilbert space and that $T: D(T) \subset E \to E$ is a linear closed operator in E. Let $\phi \in E$ and let (ϕ_n) be a sequence in E such that $\phi_n \to \phi$ in E and $\sup_{n \in \mathbb{N}} |T\phi_n|_E < \infty$. Then $\phi \in D(T)$.*

Proof. Since E is reflexive there exists a subsequence (ϕ_{n_k}) of (ϕ_n) weakly convergent to some element $\psi \in E$. Thus we have

$$\phi_{n_k} \to \phi, \quad T\phi_{n_k} \to \psi \quad \text{weakly},$$

as $k \to \infty$. Since the graph of a closed operator is also closed in the weak topology of $E \times E$, see e.g. [26], it follows that $\phi \in D(T)$ and $T\phi = \psi$. $\square$

Proposition 10.11 *We have $\mathrm{Lip}(H) \subset W^{1,2}(H, \mu)$.*

Proof. Step 1. We assume that H is N-dimensional, $N \in \mathbb{N}$.

For any $\varphi \in C_b(H)$ define,

$$\varphi_n(x) = \int_{\mathbb{R}^N} \varphi(x - y) N_{\frac{1}{n} I_N}(dy) = \left(\frac{n}{2\pi}\right)^{N/2} \int_{\mathbb{R}^N} e^{-\frac{n}{2}|x-y|^2} \varphi(y) dy.$$

Then

$$\lim_{n \to \infty} \varphi_n(x) = \varphi(x), \quad x \in H.$$

It is clear that φ_n is continuously differentiable and that, since

$$|\varphi_n(x) - \varphi_n(x_1)| \leq [\varphi]_1 |x - x_1|, \quad x, x_1 \in H,$$

we have $\|D\varphi_n\|_0 \leq [\varphi]_1$ for all $n \in \mathbb{N}$. Moreover, since

$$|\varphi(x)| \leq |\varphi(0)| + [\varphi]_1 |x|, \quad x \in H,$$

we have

$$|\varphi_n(x)| \leq |\varphi(0)| + \left(\frac{n}{2\pi}\right)^{N/2} [\varphi]_1 \int_{\mathbb{R}^N} e^{-\frac{n}{2}|x-y|^2} |y| dy, \quad x \in H,$$

so that there exists $C_1 > 0$ such that

$$|\varphi_n(x)| \leq C_1(1 + |x|), \quad x \in H.$$

Therefore, $(\varphi_n) \subset W^{1,2}(H, \mu)$ (see Exercise 10.7) and we have

$$\sup_{n \in \mathbb{N}} \left[\int_{\mathbb{R}^N} |\varphi_n|^2 d\mu + \int_{\mathbb{R}^N} |D\varphi_n|^2 d\mu \right] < \infty.$$

Therefore $\varphi \in W^{1,2}(H, \mu)$ by Proposition 10.10.

Step 2. H infinite dimensional.

For any $k \in K$ set $\varphi_k(x) = \varphi(P_k x)$. Then $\varphi_k \in W^{1,2}(H, \mu)$ for all $k \in H$ by the previous step. Since

$$\varphi_k(x) \to \varphi(x), \quad D\varphi_k(x) = P_k D\varphi(P_k x), \quad x \in H,$$

we have

$$\sup_{k \in \mathbb{N}} \left[\int_{\mathbb{R}^N} |\varphi_k|^2 d\mu + \int_{\mathbb{R}^N} |D\varphi_k|^2 d\mu \right] < \infty$$

and so, $\varphi \in W^{1,2}(H, \mu)$ again by Proposition 10.10. $\square$

10.1.5 Regularity properties of functions of $W^{1,2}(H, \mu)$

The Sobolev embedding theorem does not hold for $W^{1,2}(H, \mu)$. If $\varphi \in W^{1,2}(H, \mu)$ we can only say that $\varphi \log(|\varphi|)$ belongs to $L^2(H, \mu)$, see section 10.5 below. We want to prove here another regularity result, namely that if $\varphi \in W^{1,2}(H, \mu)$ the mapping $x \to |x|\varphi(x)$ belongs to $L^2(H, \mu)$.

Proposition 10.12 *Let $\varphi \in W^{1,2}(H, \mu)$ and let $z \in H$. Then $W_z \varphi \in L^2(H, \mu)$ and the following estimate holds.*

$$\int_H (W_z \varphi)^2 d\mu \le 2|z|^2 \int_H \varphi^2 d\mu + 4 \int_H |\langle \overline{D}\varphi, Q^{1/2}z \rangle|^2 d\mu. \qquad (10.12)$$

Proof. We first assume that $\varphi \in \mathscr{E}(H)$. Then by (10.5) we have that for any $\varphi, \psi \in \mathscr{E}(H)$ and any $z \in H$,

$$\int_H \langle D\varphi, Q^{1/2}z \rangle \, \psi \, d\mu = - \int_H \langle D\psi, Q^{1/2}z \rangle \, \varphi \, d\mu + \int_H W_z \, \varphi \, \psi \, d\mu. \qquad (10.13)$$

It is easy to see, by a standard approximation, that (10.13) holds for any $\psi \in W^{1,2}(H, \mu)$.

Now, set in (10.13) $\psi = W_{P_n z}\varphi$ (obviously $\psi \in W^{1,2}(H, \mu)$). Since

$$D(W_{P_n z}\varphi) = W_{P_n z}D\varphi + Q^{-1/2}P_n z \, \varphi,$$

we obtain

$$\int_H \langle D\varphi, Q^{1/2}z \rangle \, W_{P_n z}\varphi \, d\mu = - \int_H \langle D(W_{P_n z}\varphi), Q^{1/2}z \rangle \, \varphi \, d\mu + \int_H W_z \, W_{P_n z} \, \varphi^2 \, d\mu$$

$$= - \int_H \langle D\varphi, Q^{1/2}z \rangle \, W_{P_n z}\varphi \, d\mu - \int_H \langle P_n z, z \rangle \, \varphi^2 \, d\mu$$

$$+ \int_H W_z \, W_{P_n z}\varphi^2 \, d\mu.$$

Letting $n \to \infty$ we deduce that

$$\int_H (W_z \, \varphi)^2 \, d\mu = 2 \int_H \langle D\varphi, Q^{1/2} z \rangle \, W_z \, \varphi \, d\mu + \int_H |z|^2 \varphi^2 \, d\mu.$$

Consequently [4]

$$\int_H (W_z \, \varphi)^2 \, d\mu \le \frac{1}{2} \int_H (W_z \, \varphi)^2 \, d\mu + 2 \int_H |\langle D\varphi, Q^{1/2} z \rangle|^2 d\mu$$

$$+ \int_H |z|^2 \varphi^2 \, d\mu,$$

which yields

$$\int_H (W_z \, \varphi)^2 \, d\mu \le 4 \int_H |\langle D\varphi, Q^{1/2} z \rangle|^2 d\mu + 2 \int_H |z|^2 \varphi^2 \, d\mu. \qquad (10.14)$$

So, estimate (10.12) is proved when $\varphi \in \mathscr{E}(H)$. Assume now that $\varphi \in W^{1,2}(H, \mu)$ and let $(\varphi_n) \subset \mathscr{E}(H)$ be a sequence such that

$$\lim_{n \to \infty} \varphi_n = \varphi \quad \text{in } L^2(H, \mu), \qquad \lim_{n \to \infty} D\varphi_n = \overline{D}\varphi \quad \text{in } L^2(H, \mu; H).$$

Then by (10.14) it follows that

$$\int_H [W_z(\varphi_m - \varphi_n)]^2 \, d\mu \le 4 \int_H |\langle D(\varphi_m - \varphi_n), Q^{1/2} z \rangle|^2 d\mu$$

$$+ 2 \int_H |z|^2 (\varphi_m - \varphi_n)^2 \, d\mu. \qquad (10.15)$$

Therefore, $(W_z \varphi_n)$ is Cauchy and so, $W_z \varphi \in L^2(H, \mu)$ and (10.12) follows. $\square$

Exercise 10.13 Prove that if $\varphi \in W^{1,2}(H, \mu)$ then $|x| \varphi \in L^2(H, \mu)$.

10.2 Expansions in Wiener chaos

In this section we want to give a characterization of the space $W^{1,2}(H, \mu)$ in terms of the orthonormal system $(H_\gamma)_{\gamma \in \Gamma}$ introduced in Chapter 9.

Notice first that for any $h \in \mathbb{N}$ and any $\gamma \in \Gamma$ we have

$$D_h H_\gamma(x) = \sqrt{\frac{\gamma_h}{\lambda_h}} \, H_{\gamma^{(h)}}(x), \qquad x \in H, \ h \in \mathbb{N} \qquad (10.16)$$

---

[4] By the obvious inequality: $ab \le \frac{1}{2}(a^2 + b^2)$.

where $\gamma^{(h)} = 0$ if $\gamma_h = 0$ and if $\gamma_h > 0$,

$$\gamma_n^{(h)} = \begin{cases} \gamma_n - 1 & \text{if } h = n \\ \gamma_n & \text{if } h \neq n. \end{cases} \tag{10.17}$$

For any $\varphi \in L^2(H, \mu)$ we write as before,

$$\varphi = \sum_{\gamma \in \Gamma} \varphi_\gamma H_\gamma,$$

where

$$\varphi_\gamma = \int_H \varphi \, H_\gamma \, d\mu,$$

so that, by the Parseval identity, we have

$$\int_H \varphi^2 d\mu = \sum_{\gamma \in \Gamma} |\varphi_\gamma|^2.$$

Lemma 10.14 *Let* $\varphi \in C_b^1(H)$. *Then for any* $h \in \mathbb{N}$ *we have*

$$D_h \varphi = \sum_{\gamma \in \Gamma} \sqrt{\frac{\gamma_h}{\lambda_h}} \, \varphi_\gamma H_{\gamma^{(h)}} \tag{10.18}$$

and

$$\int_H |D_h \varphi|^2 d\mu = \sum_{\gamma \in \Gamma} \frac{\gamma_h}{\lambda_h} |\varphi_\gamma|^2. \tag{10.19}$$

Proof. Let $h \in \mathbb{N}$. It is enough to check that

$$\int_H D_h \varphi H_{\gamma^{(h)}} d\mu = \sqrt{\frac{\gamma_h}{\lambda_h}} \, \varphi_\gamma. \tag{10.20}$$

We assume for simplicity that $\gamma_h \geq 2$. By (3.1) we have

$$\int_H D_h \varphi H_{\gamma^{(h)}} d\mu = -\int_H \varphi \, D_h H_{\gamma^{(h)}} d\mu + \int_H \frac{x_h}{\lambda_h} H_{\gamma^{(h)}} \varphi d\mu$$

$$= \int_H \varphi(x) \prod_{j \neq h} H_{\gamma_j} \left(\frac{x_j}{\sqrt{\lambda_j}} \right)$$

$$\times \left(\frac{x_h}{\lambda_h} H_{\gamma_h - 1} \left(\frac{x_h}{\sqrt{\lambda_h}} \right) - \sqrt{\frac{\gamma_h - 1}{\lambda_h}} H_{\gamma_h - 2} \left(\frac{x_k}{\sqrt{\lambda_k}} \right) \right) d\mu.$$

Now, taking into account (9.7) we find,

$$\frac{x_h}{\lambda_h} H_{\gamma_h - 1} \left(\frac{x_h}{\sqrt{\lambda_h}} \right) = \sqrt{\frac{\gamma_h}{\lambda_h}} H_{\gamma_h} \left(\frac{x_h}{\sqrt{\lambda_h}} \right) + \sqrt{\frac{\gamma_h - 1}{\lambda_h}} H_{\gamma_h - 2} \left(\frac{x_k}{\sqrt{\lambda_k}} \right),$$

which yields (10.20) and consequently (10.18). Finally, (10.19) follows from the Parseval identity. $\square$

Theorem 10.15 *Assume that $\varphi \in W^{1,2}(H, \mu)$. Then for any $h \in \mathbb{N}$ we have*

$$\overline{D}_h \varphi = \sum_{\gamma \in \Gamma} \sqrt{\frac{\gamma_h}{\lambda_h}}\, \varphi_\gamma H_{\gamma^{(h)}} \tag{10.21}$$

and

$$\int_H |\overline{D}\varphi|^2 d\mu = \sum_{\gamma \in \Gamma} \langle \gamma, \lambda^{-1}\rangle |\varphi_\gamma|^2, \tag{10.22}$$

where $\langle \gamma, \lambda^{-1}\rangle = \sum_{h=1}^{\infty} \dfrac{\gamma_h}{\lambda_h}$. Conversely if

$$\sum_{\gamma \in \Gamma} \langle \gamma, \lambda^{-1}\rangle |\varphi_\gamma|^2 < +\infty, \tag{10.23}$$

then $\varphi \in W^{1,2}(H, \mu)$.

Proof. Let $\varphi \in W^{1,2}(H, \mu)$ and let $(\varphi_n) \subset \mathscr{E}(H)$ be such that

$$\varphi_n \to \varphi \quad \text{in } L^2(H, \mu), \quad D\varphi_n \to \overline{D}\varphi \quad \text{in } L^2(H, \mu; H).$$

By Lemma 10.14 it follows that for any $h \in \mathbb{N}$,

$$\int_{II} D_h\varphi H_{\gamma^{(h)}} d\mu = \lim_{n \to \infty} \int_{II} D_h\varphi_n H_{\gamma^{(h)}} d\mu = \sqrt{\frac{\gamma_h}{\lambda_h}}\, \varphi_\gamma,$$

so that (10.21), and consequently (10.22), holds.

Assume conversely that (10.23) holds. Write

$$\Gamma_N := \{\gamma \in \Gamma : \langle \gamma, \lambda^{-1}\rangle \le N\}, \quad N \in \mathbb{N}.$$

Since $\lambda_n \to 0$ as $n \to \infty$, each set Γ_N is finite, moreover,

$$\Gamma_N \uparrow \Gamma.$$

For any $N \in \mathbb{N}$ set

$$\varphi^{(N)} = \sum_{\gamma \in \Gamma_N} \varphi_\gamma H_\gamma,$$

it is clear that $\varphi^{(N)} \in W^{1,2}(H, \mu)$ and that

$$\lim_{N \to \infty} \varphi^{(N)} \to \varphi \quad \text{in } L^2(H, \mu).$$

Moreover, if $N, p \in \mathbb{N}$ we have

$$\int_H |D\varphi^{(N+p)} - D\varphi^{(N)}|^2 d\mu = \sum_{\gamma \in \Gamma_{N+p} \setminus \Gamma_N} \langle \gamma, \lambda^{-1} \rangle |\varphi_\gamma|^2.$$

By (10.23) it follows that $(D\varphi^{(N)})$ is Cauchy in $L^2(H, \mu; H)$ and this implies that φ belongs to $W^{1,2}(H, \mu)$ as required. $\square$

10.2.1 Compactness of the embedding of $W^{1,2}(H, \mu)$ in $L^2(H, \mu)$

Theorem 10.16 *The embedding* $W^{1,2}(H, \mu) \subset L^2(H, \mu)$ *is compact.*

Proof. Let $(\varphi^{(n)})$ be a sequence in $W^{1,2}(H, \mu)$ such that

$$\int_H \left[|\varphi^{(n)}|^2 + |D\varphi^{(n)}|^2 \right] d\mu \le K,$$

where $K > 0$. We have to show that there exists a subsequence of $(\varphi^{(n)})$ convergent in $L^2(H, \mu)$.

Since $L^2(H, \mu)$ is reflexive, there exists a subsequence $(\varphi^{(n_k)})$ of $(\varphi^{(n)})$ weakly convergent to some function $\varphi \in L^2(H, \mu)$. We are going to show that $\varphi^{(n_k)}$ converges indeed to φ strongly.

Write for any $M \in \mathbb{N}$,

$$\int_H |\varphi - \varphi^{(n_k)}|^2 d\mu = \sum_{\gamma \in \Gamma_N} |\varphi_\gamma - \varphi_\gamma^{(n_k)}|^2 + \sum_{\gamma \in (\Gamma_N)^c} |\varphi_\gamma - \varphi_\gamma^{(n_k)}|^2$$

$$\le \sum_{\gamma \in \Gamma_N} |\varphi_\gamma - \varphi_\gamma^{(n_k)}|^2 + \frac{1}{N} \sum_{\gamma \in \Gamma} \langle \gamma, \lambda^{-1} \rangle |\varphi_\gamma - \varphi_\gamma^{(n_k)}|^2. \qquad (10.24)$$

By Theorem 10.15, it follows that for any $k \in K$

$$\sum_{\gamma \in \Gamma} \langle \gamma, \lambda^{-1} \rangle |\varphi_\gamma^{(n_k)}|^2 \le K,$$

which implies that

$$\sum_{\gamma \in \Gamma} \langle \gamma, \lambda^{-1} \rangle |\varphi_\gamma|^2 \le K.$$

Consequently, by (10.24) we have

$$\int_H |\varphi - \varphi^{(n_k)}|^2 d\mu \le \sum_{\gamma \in \Gamma_N} |\varphi_\gamma - \varphi_\gamma^{(n_k)}|^2 + \frac{2K}{M}. \qquad (10.25)$$

Since the set Γ_N is finite and $(\varphi^{(n_k)})$ is weakly convergent to φ we have

$$\lim_{n\to\infty} \sum_{\gamma\in\Gamma_N} |\varphi_\gamma - \varphi_\gamma^{(n)}|^2 = 0,$$

and so the conclusion follows by the arbitrariness of N and taking into account (10.25). $\square$

10.3 The adjoint of $\overline{D}$

10.3.1 Adjoint operator

Let us first recall the definition of *adjoint* operator. Let E, F be Hilbert spaces and let $T\colon D(T) \subset E \to F$ be a linear closed operator with domain $D(T)$ dense in E. For any $y \in F$ consider the linear functional

$$\Lambda_y(x) = \langle Tx, y\rangle_F, \quad x \in D(T).$$

Define

$$D(T^*) = \{y \in F : \Lambda_y \text{ is continuous}\}.$$

By the Riesz representation theorem for any $y \in D(T^*)$ there exists a unique element $z \in E$ such that

$$\Lambda_y(x) = \langle x, z\rangle_E \quad \text{for all } x \in E.$$

We set $T^*y = z$. It is easy to see that T^* is a closed operator in F.

Remark 10.17 *If Y is a core* [5] *for T then the previous considerations hold if we define Λ on Y instead of in $D(T)$.*

10.3.2 The adjoint operator of $\overline{D}$

We shall denote by D^* the adjoint operator of $\overline{D}$.

Let us first consider the case when H has finite dimension n.

Proposition 10.18 *Let H be n-dimensional and let $F \in C_b^1(H; H)$. Then F belongs to the domain of D^* and we have*

$$D^*F(x) = -\text{div } F(x) + \langle Q^{-1}x, F(x)\rangle, \quad x \in H, \tag{10.26}$$

where

$$\text{div } F(x) = \sum_{h=1}^{n} D_h F_h(x), \quad x \in H$$

------

[5] See Appendix A.3.1.

and

$$F_h(x) = \langle F(x), e_h \rangle, \quad x \in H.$$

Proof. Write

$$F(x) = \sum_{h=1}^{n} F_h(x)e_h, \quad x \in H,$$

and let $\varphi \in \mathscr{E}(H)$. Then by (10.1) with $\psi = F_h$ and $z = e_h$ we have

$$\int_H \langle D\varphi, e_h \rangle F_h \, d\mu = - \int_H \langle DF_h, e_h \rangle \varphi \, d\mu + \int_H \langle Q^{-1}x, e_h \rangle F_h\varphi \, d\mu.$$

Summing up on h yields

$$\Lambda_F(\varphi) := \int_H \langle D\varphi, F \rangle \, d\mu = - \int_H \operatorname{div} F \, \varphi \, d\mu + \int_H \langle Q^{-1}x, F \rangle \, \varphi \, d\mu.$$

It follows that

$$|\Lambda_F(\varphi)| \leq \left[\left(\int_H |\operatorname{div} F|^2 d\mu \right)^{1/2} + \left(\int_H |\langle Q^{-1}x, F \rangle|^2 d\mu \right)^{1/2} \right]$$

$$\times \left(\int_H |\varphi|^2) d\mu \right)^{1/2},$$

for all $\varphi \in \mathscr{E}(H)$. Since $\mathscr{E}(H)$ is a core for $\overline{D}$ it follows that F belongs to the domain of D^* and (10.26) holds. $\square$

Exercise 10.19 Prove that for any $k \in \mathbb{N}$ and any $\varphi \in W^{1,2}(H, \mu)$ we have

$$D_k^*\varphi = -D_k\varphi + \frac{x_k}{\lambda_k}\,\varphi.$$

In the infinite dimensional case we are not able to give a meaning to formula (10.26) in general, but only for special vector fields F.

Proposition 10.20 *Let $\psi \in W^{1,2}(H, \mu)$, $z \in Q^{1/2}(H)$ and let F be given by*

$$F(x) = \psi(x)z, \quad x \in H.$$

Then F belongs to the domain of D^* and we have

$$D^*F(x) = -\langle \overline{D}\psi(x), z \rangle + W_{Q^{-1/2}z}(x)\psi(x), \quad x \in H. \tag{10.27}$$

Proof. Let $\varphi \in \mathscr{E}(H)$. Then by (10.13) we have

$$\Lambda_F(\varphi) = \int_H \langle D\varphi, z \rangle\, \psi\, d\mu = -\int_H \langle \overline{D}\psi, z \rangle\, \varphi\, d\mu + \int_H W_{Q^{-1/2}z}\, \varphi\, \psi\, d\mu,$$

so that Λ_F is continuous and the conclusion follows. $\square$

Proposition 10.21 *Assume that $\psi_0 \in W^{1,2}(H, \mu)$ and that F belongs to the domain of D^*. Assume that ψ_0 and $\overline{D}\psi_0$ are bounded. Then $\psi_0 F$ belongs to the domain of D^* and*

$$D^*(\psi_0 F) = \psi_0 \overline{D}^*(F) - \langle \overline{D}\psi_0, F \rangle. \tag{10.28}$$

Proof. For any $\varphi \in \mathscr{E}(H)$ we have

$$\Lambda_{\psi_0 F}(\varphi) = \int_H \langle D\varphi, F \rangle\, \psi_0\, d\mu = \int_H \langle \psi_0 D\varphi, F \rangle\, d\mu.$$

Using (10.11) we can write

$$\Lambda_{\psi_0 F}(\varphi) = \int_H \langle D(\psi_0 \varphi), F \rangle\, d\mu - \int_H \langle D\psi_0, F \rangle\, \varphi\, d\mu$$

$$= \int_H \varphi \psi_0 D^* F\, d\mu - \int_H \langle D\psi_0, F \rangle\, \varphi\, d\mu.$$

and the conclusion follows. $\square$

10.4 The Dirichlet form associated to μ

In this section we write $\overline{D} = D$ for simplicity.

Let us consider the bilinear form,

$$a: W^{1,2}(H, \mu) \times W^{1,2}(H, \mu) \to \mathbb{R},$$

$$(\varphi, \psi) \mapsto a(\varphi, \psi) = \frac{1}{2} \int_H \langle D\varphi, D\psi \rangle d\mu.$$

Clearly a is continuous, symmetric and coercive. Therefore, by the Lax–Milgram theorem there exists a unique negative self-adjoint operator

$$L_2: D(L_2) \subset L^2(H, \mu) \to L^2(H, \mu)$$

such that

$$a(\varphi, \psi) = -\int_H L_2\varphi\, \psi\, d\mu = -\int_H \varphi\, L_2\psi\, d\mu.$$

We want now to identify the operator L_2. For this it is useful to introduce the negative self-adjoint operator $A := -\frac{1}{2} Q^{-1}$ and to consider the subspace $\mathscr{E}_A(H)$ of $\mathscr{E}(H)$ defined by (8.28).

Let $\varphi, \psi \in \mathscr{E}_A(H)$. Then by the integration by parts formula (10.1) it follows that

$$\int_H \langle D\varphi, D\psi \rangle d\mu = \sum_{k=1}^{\infty} \int_H D_k\varphi \, D_k\psi \, d\mu$$

$$= -\sum_{k=1}^{\infty} \int_H \varphi D_k^2 \, \psi \, d\mu + \sum_{k=1}^{\infty} \frac{1}{\lambda_k} \int_H x_k\varphi \, D_k\psi \, d\mu$$

$$= -\int_H \varphi \, \mathrm{Tr} \, [D^2\psi] d\mu + \int_H \langle x, Q^{-1}D\psi \rangle \varphi d\mu.$$

Notice that the term $Q^{-1}D\psi$ is meaningful since $\psi \in \mathscr{E}_A(H)$ (this is the reason for choosing $\mathscr{E}_A(H)$ instead of $\mathscr{E}(H)$).

Therefore it follows that

$$L_2\varphi = \frac{1}{2} \, \mathrm{Tr} \, [D^2\varphi] + \langle x, AD\varphi \rangle, \quad \varphi \in \mathscr{E}_A(H). \tag{10.29}$$

This fact leads us to introduce the following Ornstein–Uhlenbeck semigroup

$$R_t\varphi(x) = \int_H \varphi(e^{tA}x + y)N_{Q_t}(dy), \quad \varphi \in B_b(H), \tag{10.30}$$

where

$$Q_t x = \int_0^t e^{2sA}x ds = Q(1 - e^{2tA})x, \quad x \in H. \tag{10.31}$$

In the next proposition we list several properties of R_t.

Proposition 10.22 *The following statements hold.*

(i) R_t is strong Feller.

(ii) μ is the unique invariant measure of R_t. Moreover, it is ergodic and strongly mixing,

$$\lim_{t \to +\infty} R_t\varphi(x) = \int_H \varphi(y)\mu(dy) := \overline{\varphi} \quad in \, L^2(H, \mu). \tag{10.32}$$

(iii) R_t can be uniquely extended to a strongly continuous semigroup of contractions in $L^2(H, \mu)$ (which we still denote by R_t).

(iv) $\mathscr{E}_A(H)$ is a core for R_t and the infinitesimal generator of R_t is precisely the operator L_2 defined by (10.30).

(v) R_t is symmetric, so that its infinitesimal generator L_2 is self-adjoint in $L^2(H, \mu)$.

Proof. (i) First notice that the controllability condition (8.18) is fulfilled since $C = I$, see Remark 8.15. Now the conclusion follows from Theorem 8.16.

Moreover (ii) follows from Theorem 8.20, (iii) from Proposition 8.21 and (iv) from Theorem 8.22.

Let us finally show (v) which is equivalent to

$$\int_H R_t\varphi\,\psi\,d\mu = \int_H \varphi\, R_t\psi\,d\mu, \quad \varphi, \psi \in L^2(H, \mu). \tag{10.33}$$

Since $\mathscr{E}_A(H)$ is dense in $L^2(H, \mu)$, it is enough to prove (10.33) for $\varphi, \psi \in \mathscr{E}_A(H)$. Let $\varphi = \varphi_h, \psi = \varphi_k$ with $h, k \in D(A)$. Then we have

$$\int_H R_t\varphi\,\psi\,d\mu = \int_H e^{-\frac{1}{2}\langle Q_t h, h\rangle} e^{i\langle x, e^{tA}h\rangle} e^{i\langle x, k\rangle}\,d\mu$$

$$= e^{-\frac{1}{2}\langle(Q_t + e^{tA}Q_\infty e^{tA})h, h\rangle} e^{-\frac{1}{2}[\langle Q_\infty e^{tA}h, k\rangle + \langle Q_\infty e^{tA}k, h\rangle]}$$

$$= e^{-\frac{1}{2}\langle(Q_\infty h, h\rangle} e^{-\frac{1}{2}[\langle Q_\infty e^{tA}h, k\rangle + \langle Q_\infty e^{tA}k, h\rangle]}$$

$$= \int_H \varphi\, R_t\psi\,d\mu.$$

The proof is complete. $\square$

Proposition 10.23 We have $D(L_2) \subset W^{1,2}(H, \mu)$ with continuous and dense embedding. Moreover for all $\varphi \in D(L_2)$ we have

$$\int_H L_2\varphi\,\varphi\,d\mu = -\frac{1}{2}\int_H |D\varphi|^2 d\mu \tag{10.34}$$

and for all $\varphi, \psi \in D(L_2)$

$$\int_H L_2\varphi\,\psi\,d\mu = -\frac{1}{2}\int_H \langle D\varphi, D\psi\rangle d\mu = -a(\varphi, \psi). \tag{10.35}$$

Proof. We first prove (10.34) when $\varphi \in \mathscr{E}_A(H)$. For this let us first show the following identity,

$$L_2(\varphi^2) = 2\varphi L_2\varphi + |D\varphi|^2, \quad \varphi \in \mathscr{E}_A(H). \tag{10.36}$$

We have in fact

$$D(\varphi^2) = 2\varphi D\varphi, \quad D^2(\varphi^2) = 2\varphi D^2\varphi + 2D\varphi \otimes D\varphi,$$

which implies that

$$\mathrm{Tr}\,[D^2(\varphi^2)] = 2\varphi\,\mathrm{Tr}\,[D^2\varphi] + 2|D\varphi|^2.$$

Consequently, since $\varphi^2 \in \mathscr{E}_A(H)$, we have

$$L_2(\varphi^2) = \varphi\,\mathrm{Tr}\,[D^2\varphi] + |D\varphi|^2 + 2\varphi\langle x, AD\varphi\rangle$$

and so (10.36) holds.

Now integrating both sides of (10.36) with respect to μ, and taking into account that $\int_H L_2(\varphi^2)d\mu = 0$ (because μ is invariant), yields (10.34) when $\varphi \in \mathscr{E}_A(H)$.

Let now $\varphi \in D(L_2)$. Since $\mathscr{E}_A(H)$ is a core there exists a sequence $(\varphi_n) \subset \mathscr{E}_A(H)$ such that

$$\varphi_n \to \varphi, \quad L_2\varphi_n \to L_2\varphi \quad \text{in } L^2(H, \mu).$$

By (10.34) it follows that for $n, m \in \mathbb{N}$,

$$\int_H |D\varphi_n - D\varphi_m|^2 d\mu = -2\int_H L_2(\varphi_n - \varphi_m)(\varphi_n - \varphi_m)\,d\mu.$$

This implies that the sequence (φ_n) is Cauchy in $W^{1,2}(H, \mu)$, and so, $\varphi \in W^{1,2}(H, \mu)$ as claimed. Finally, (10.35) can be proved similarly, we leave the proof to the reader. $\square$

We end this subsection by giving another interesting proof of Proposition 10.9.

Proposition 10.24 *Assume that $\varphi : H \to \mathbb{R}$ is Lipschitz continuous. Then $\varphi \in W^{1,2}(H, \mu)$.*

Proof. For any $t > 0$ we have $R_t\varphi \in C_b^1(H)$ (by Theorem 8.16) and for any $h \in \mathbb{N}$ we have,

$$D_h R_t\varphi(x) = \int_H \langle \Gamma(t)e_h, Q_t^{-1/2}y\rangle \varphi(e^{tA}x + y)N_{Q_t}(dy),$$

where $\Gamma(t) = Q_t^{-1/2}e^{tA}$. It follows that

$$|D_h R_t\varphi(x)|^2 \le \frac{1}{t}R_t(\varphi^2)(x), \quad x \in H, \ t \ge 0,$$

which implies that

$$\int_H |D_h R_t \varphi(x)|^2 d\mu \le \frac{1}{t} \int_H \varphi^2 d\mu.$$

Consequently there is a sequence $t_n \to +\infty$ such that $D_h R_t \varphi$ converges weakly to some element $g \in L^2(H, \mu)$. This implies that $D_h \varphi = g_h$, and, due to the arbitrariness of h, that $\varphi \in W^{1,2}(H, \mu)$. $\square$

10.5 Poincaré and log-Sobolev inequalities

Let us fix some notations. Setting $\omega = \frac{1}{2} \inf_{k \in \mathbb{N}} \frac{1}{\lambda_k}$, we have clearly $\omega = \|Q\|^{-1} > 0$. Moreover, since

$$e^{tA} x = \sum_{k=1}^{\infty} e^{-\frac{t}{2\lambda_k}} \langle x, e_k \rangle e_k, \quad x \in H,$$

it follows by the Parseval identity that

$$|e^{tA} x|^2 = \sum_{k=1}^{\infty} e^{-\frac{t}{\lambda_k}} |\langle x, e_k \rangle|^2, \quad x \in H,$$

and consequently

$$\|e^{tA}\| \le e^{-\omega t}, \quad t \ge 0, \tag{10.37}$$

We start with the *Poincaré inequality*.

Theorem 10.25 *For all $\varphi \in W^{1,2}(H, \mu)$ we have*

$$\int_H |\varphi - \overline{\varphi}|^2 d\mu \le \frac{1}{2\omega} \int_H |D\varphi|^2 d\mu, \tag{10.38}$$

where

$$\overline{\varphi} = \int_H \varphi d\mu.$$

Proof. Thanks to Proposition 10.23, it is enough to prove (10.38) for $\varphi \in D(L_2)$. In this case we have, by Proposition A.7,

$$\frac{d}{dt} R_t \varphi = L_2 R_t \varphi, \quad t \ge 0.$$

Multiplying both sides of this identity by $R_t \varphi$, integrating with respect to μ over H and taking into account (10.34), we find

$$\frac{1}{2} \frac{d}{dt} \int_H |R_t \varphi|^2 d\mu = \int_H L_2 R_t \varphi \, R_t \varphi \, d\mu = -\frac{1}{2} \int_H |D R_t \varphi|^2 d\mu. \tag{10.39}$$

Now we need a suitable estimate for $|DR_t\varphi|^2$. First we notice that for any $h \in H$ we have

$$\langle DR_t\varphi(x), h\rangle = \int_H \langle D\varphi(e^{tA}x + y), e^{tA}h\rangle N_{Q_t}(dy), \quad x \in H.$$

It follows, by using the Hölder inequality and (10.37), that

$$|\langle DR_t\varphi(x), h\rangle|^2 \le e^{-2\omega t} \int_H |D\varphi(e^{tA}x + y)|^2 N_{Q_t}(dy)$$

$$= e^{-2\omega t} R_t(|D\varphi|^2) \, |h|^2, \quad x, h \in H.$$

The arbitrariness of h yields

$$|DR_t\varphi(x)|^2 \le e^{-2\omega t} R_t(|D\varphi|^2)(x), \quad x \in H.$$

Now, substituting $|DR_t\varphi(x)|^2$ in (10.39) and taking into account the invariance of the measure μ, we obtain

$$\frac{d}{dt} \int_H |R_t\varphi|^2 d\mu \ge -e^{-2\omega t} \int_H R_t(|D\varphi|^2) d\mu = -e^{-2\omega t} \int_H |D\varphi|^2 d\mu.$$

Integrating in t yields

$$\int_H |R_t\varphi|^2 d\mu - \int_H \varphi^2 d\mu \ge -\frac{1}{2\omega}(1 - e^{-2\omega t}) \int_H |D\varphi|^2 d\mu.$$

Finally, letting t tend to $+\infty$, and recalling (10.32), yields

$$(\overline{\varphi})^2 - \int_H \varphi^2 d\mu \ge -\frac{1}{2\omega} \int_H |D\varphi|^2 d\mu$$

and the conclusion follows, since

$$\int_H |\varphi - \overline{\varphi}|^2 d\mu = \int_H \varphi^2 d\mu - (\overline{\varphi})^2.$$

$\square$

A first immediate consequence of the Poincaré inequality is the following.

Proposition 10.26 Let $\varphi \in W^{1,2}(H, \mu)$ such that $D\varphi = 0$. Then $\varphi = \overline{\varphi}$.

Next we show that the spectrum $\sigma(L_2)$ of L_2 consists of 0 and a set included in a half-space $\{\lambda \in \mathbb{C} : \Re\lambda \le -\omega\}$ (spectral gap). We have in fact the following result.

Proposition 10.27 *We have*

$$\sigma(L_2) \backslash \{0\} \subset \{\lambda \in \mathbb{C} : \mathfrak{Re} \, \lambda \leq -\omega\}. \tag{10.40}$$

Proof. Let us consider the space

$$L^2_\pi(H, \mu) := \left\{ \varphi \in L^2(H, \mu) : \overline{\varphi} = 0 \right\}.$$

Clearly

$$L^2(H, \mu) = L^2_\pi(H, \mu) \oplus \mathbb{R}.$$

Moreover if $\overline{\varphi} = 0$ we have

$$\overline{(R_t \varphi)} = \int_H R_t \varphi d\mu = \int_H \varphi d\mu = 0,$$

so that $L^2_\pi(H, \mu)$ is an invariant subspace of R_t. Denote by L_π the restriction of L_2 to $L^2_\pi(H, \mu)$. Then we have clearly

$$\sigma(L_2) = \{0\} \cup \sigma(L_\pi).$$

Moreover if $\varphi \in L^2_\pi(H, \mu)$ we see, using the Poincaré inequality, that

$$\int_H L_\pi \varphi \, \varphi \, d\mu = -\frac{1}{2} \int_H |D\varphi|^2 d\mu \leq -\omega \int_H \varphi^2 d\mu, \tag{10.41}$$

which yields

$$\sigma(L_\pi) \subset \{\lambda \in \mathbb{C} : \mathrm{Re} \, \lambda \leq -\omega\}. \quad \square$$

The spectral gap implies the exponential convergence of $R_t \varphi$ to the equilibrium, as the following proposition shows.

Proposition 10.28 *For any $\varphi \in L^2(H, \mu)$ we have*

$$\int_H |R_t \varphi - \overline{\varphi}|^2 d\mu \leq e^{-\omega t} \int_H |\varphi|^2 d\mu, \quad \varphi \in L^2(H, \mu). \tag{10.42}$$

Proof. Let first $\varphi \in L^2_\pi(H, \mu)$. Then by (10.41) and the Hille–Yosida theorem (Theorem A.15), we have

$$\int_H |R_t \varphi|^2 d\mu \leq e^{-\omega t} \int_H |\varphi|^2 d\mu, \quad \varphi \in L^2(H, \mu). \tag{10.43}$$

Now let $\varphi \in L^2(H, \mu)$ be arbitrary. Then, replacing in (10.43) φ with $\varphi - \overline{\varphi}$, we obtain

$$\int_H |R_t \varphi - \overline{\varphi}|^2 d\mu = \int_H |R_t (\varphi - \overline{\varphi})|^2 d\mu$$

$$\leq e^{-\omega t} \int_H |\varphi - \overline{\varphi}|^2 d\mu \leq e^{-\omega t} \int_H |\varphi|^2 d\mu.$$

So, (10.42) is proved. $\square$

We are going to prove the logarithmic Sobolev inequality. For this we need a lemma.

Lemma 10.29 *For any $g \in C^2(\mathbb{R})$ such that $g(\varphi) \in D(L_2)$ we have*

$$L_2(g(\varphi)) = g'(\varphi)L_2\varphi + \frac{1}{2} g''(\varphi)|D\varphi|^2, \quad \varphi \in \mathscr{E}_A(H), \quad (10.44)$$

and

$$\int_H L_2\varphi \, g'(\varphi) \, d\mu = -\frac{1}{2} \int_H g''(\varphi)|D\varphi|^2 d\mu, \quad \varphi \in \mathscr{E}_A(H). \quad (10.45)$$

Proof. Let $\varphi \in \mathscr{E}_A(H)$. Since

$$Dg(\varphi) = g'(\varphi)D\varphi, \quad D^2g(\varphi) = g''(\varphi)D\varphi \otimes D\varphi + g'(\varphi)D^2\varphi,$$

we have

$$L_2(g(\varphi)) = \frac{1}{2} g''(\varphi)|D\varphi|^2 + \frac{1}{2} g'(\varphi)\mathrm{Tr} \, [D^2\varphi] + g'(\varphi)\langle x, D\varphi\rangle.$$

So, (10.44) follows. Finally integrating (10.44) with respect to μ over H yields

$$\int_H L_2(g(\varphi))d\mu = 0,$$

by the invariance of μ. So, (10.45) is proved. $\square$

We are now ready to prove the *log-Sobolev inequality*.

Theorem 10.30 *For all $\varphi \in W^{1,2}(H, \mu)$ we have*

$$\int_H \varphi^2 \log(\varphi^2)d\mu \le \frac{1}{\omega} \int_H |D\varphi|^2 d\mu + \|\varphi\|^2_{L^2(H,\mu)} \log(\|\varphi\|^2_{L^2(H,\mu)}).$$
$$(10.46)$$

Proof. It is enough to prove the result when $\varphi \in \mathscr{E}_A(H)$ is such that $\varphi(x) \ge \varepsilon > 0$, $x \in H$. In this case we have

$$\frac{d}{dt} \int_H (R_t(\varphi^2)) \log(R_t(\varphi^2))d\mu = \int_H L_2 R_t(\varphi^2) \log(R_t(\varphi^2))d\mu$$
$$+ \int_H L_2 R_t(\varphi^2)d\mu.$$

Now the second term in the right-hand side vanishes, due to the invariance of μ. For the first term we use (10.45) with $g'(\xi) = \log \xi$. [6] Therefore we have

_____

[6] Indeed, we should take a suitable regularization of g in order to use Lemma 10.29.

$$\frac{d}{dt} \int_H R_t(\varphi^2) \log(R_t(\varphi^2)) d\mu = -\frac{1}{2} \int_H \frac{1}{R_t(\varphi^2)} |DR_t(\varphi^2)|^2 d\mu. \quad (10.47)$$

For any $h \in H$ we have

$$\langle DR_t(\varphi^2)(x), h \rangle = 2 \int_H \varphi(e^{tA}x + y) \langle D\varphi(e^{tA}x + y), e^{tA}h \rangle N_{Q_t}(dy).$$

It follows by the Hölder inequality that

$$|\langle DR_t(\varphi^2)(x), h \rangle|^2 \le 4e^{-2t\omega} \int_H \varphi^2(e^{tA}x + y) N_{Q_t}(dy)$$

$$\times \int_H |D\varphi(e^{tA}x + y)|^2 N_{Q_t}(dy),$$

which yields

$$|DR_t(\varphi^2)|^2 \le 4e^{-2t\omega} R_t(\varphi^2) R_t(|D\varphi|^2).$$

Substituting in (10.47) gives

$$\frac{d}{dt} \int_H R_t(\varphi^2) \log(R_t(\varphi^2)) d\mu \ge -2e^{-2t\omega} \int_H R_t(|D\varphi|^2) d\mu$$

$$= -2e^{-2t\omega} \int_H |D\varphi|^2 d\mu,$$

due to the invariance of μ. Integrating in t gives

$$\int_H R_t(\varphi^2) \log(R_t(\varphi^2)) d\mu - \int_H \varphi^2 \log(\varphi^2) d\mu \ge \frac{1}{\omega} (1 - e^{-2t\omega}) \int_H |D\varphi|^2 d\mu.$$

Finally, letting t tend to $+\infty$, and recalling (10.32), yields,

$$\|\varphi\|^2_{L^2(H,\mu)} \log(\|\varphi\|^2_{L^2(H,\mu)}) - \int_H \varphi^2 \log(\varphi^2) d\mu \ge -\frac{1}{\omega} \int_H |D\varphi|^2 d\mu$$

and the conclusion follows. $\square$

10.5.1 Hypercontractivity

We consider here the semigroup R_t defined by (10.30) and its invariant measure μ. We show now that R_t is *hypercontractive*, see [16].

Theorem 10.31 *For all $t > 0$ we have*

$$\|R_t\varphi\|_{L^{q(t)}(H,\mu)} \le \|\varphi\|_{L^p(H,\mu)}, \quad p \ge 2, \ \varphi \in L^p(H,\mu), \tag{10.48}$$

where

$$q(t) = 1 + (p-1)e^{2\omega t}, \quad t > 0. \tag{10.49}$$

Proof. It is enough to show (10.48) for $\varphi \ge \varepsilon > 0$ and $\varphi \in \mathscr{E}_A(H)$. We set

$$G(t) = \|R_t\varphi\|_{L^{q(t)}(H,\mu)}, \quad F(t) = G(t)^{q(t)} = \int_H (R_t\varphi)^{q(t)} d\mu.$$

We are going to show that

$$G'(t) \le 0. \tag{10.50}$$

This will imply that $G(t) \le G(0)$, which coincides with (10.48). Since

$$G'(t) = G(t)\left(-\frac{q'(t)}{q^2(t)} \log F(t) + \frac{1}{q(t)} \frac{F'(t)}{F(t)} \right),$$

it is enough to show that

$$-\frac{1}{q(t)} F(t) \log F(t) + \frac{F'(t)}{q'(t)} \le 0. \tag{10.51}$$

Notice now that

$$F'(t) = \int_H (R_t\varphi)^{q(t)} q'(t) \log(R_t\varphi) d\mu + q(t) \int_H (R_t\varphi)^{q(t)-1} L_2 R_t\varphi d\mu. \tag{10.52}$$

Setting $f = (R_t\varphi)^{\frac{q(t)}{2}}$ and using (10.34), we find that

$$\frac{F'(t)}{q'(t)} = \int_H f^2 \log(f^{\frac{2}{q(t)}}) d\mu + \frac{q(t)}{q'(t)} \int_H f^{2\frac{q(t)-1}{q(t)}} L_2\left(f^{\frac{2}{q(t)}}\right) d\mu$$

$$= \frac{1}{q(t)} \int_H f^2 \log(f^2) d\mu - \frac{q(t)}{2q'(t)}$$

$$\times \int_H \left\langle D\left(f^{\frac{2}{q(t)}}\right), D\left(f^{2\frac{q(t)-1}{q(t)}}\right) \right\rangle d\mu$$

$$= \frac{1}{q(t)} \int_H f^2 \log(f^2) d\mu - 2\frac{q(t)-1}{q'(t)q^2(t)} \int_H |Df|^2 d\mu.$$

Consequently (10.51) is equivalent to

$$\int_H f^2 \log(f^2) d\mu \le 2\frac{q(t)-1}{q'(t)} \int_H |Df|^2 d\mu + \overline{f^2} \log(\overline{f^2}). \tag{10.53}$$

Since
$$2\,\frac{q(t) - 1}{q'(t)q(t)} = \frac{1}{\omega},$$
we see that (10.53) holds in view of the log-Sobolev inequality (10.46). $\square$

10.6 The Sobolev space $W^{2,2}(H, \mu)$

To define $W^{2,2}(H, \mu)$ we need a result that it is a generalization of Proposition 10.3. The simple proof is left to the reader.

Proposition 10.32 *For any $h, k \in \mathbb{N}$ the linear operator $D_h D_k$, defined in $\mathscr{E}(H)$, is closable.*

We shall denote by $\overline{D_h D_k}$ the closure of $D_h D_k$. If φ belongs to the domain of $\overline{D_h D_k}$ we say that $\overline{D_h D_k}\varphi$ belongs to $L^2(H, \mu)$.

Now we denote by $W^{2,2}(H, \mu)$ the linear space of all functions $\varphi \in L^2(H, \mu)$ such that $D_h D_k \varphi \in L^2(H, \mu)$ for all $h, k \in \mathbb{N}$ and

$$\sum_{h,k=1}^{\infty} \int_H |D_h D_k \varphi(x)|^2 \mu(dx) < +\infty.$$

Endowed with the inner product

$$\langle \varphi, \psi \rangle_{W^{2,2}(H,\mu)} = \langle \varphi, \psi \rangle_{L^2(H,\mu)} + \sum_{k=1}^{\infty} \int_H (\overline{D_k \varphi})(\overline{D_k \psi}) d\mu$$

$$+ \sum_{h,k=1}^{\infty} \int_H (\overline{D_h D_k \varphi}(x))(\overline{D_h D_k \psi}(x)) \mu(dx),$$

$W^{2,2}(H, \mu)$ is a Hilbert space.

If $\varphi \in W^{2,2}(H, \mu)$ we define $\overline{D^2}\varphi$ as follows

$$\langle \overline{D^2}\varphi(x)z, z \rangle = \sum_{h,k=1}^{\infty} \overline{D_h D_k \varphi}(x) z_h z_k, \quad x, z \in H, \ \mu\text{-a.s.}$$

It is easy to see that $\overline{D}^2 \varphi(x)$ is a Hilbert–Schmidt operator for almost all $x \in H$.

The following result it is a generalization of Proposition 10.12. The proof is left to the reader as an exercise.

Proposition 10.33 *If $\varphi \in W^{2,2}(H, \mu)$ then $|x|\varphi \in W^{1,2}(H, \mu)$, and $|x|^2\varphi \in L^2(H, \mu)$.*

We can now characterize the domain of the infinitesimal generator L_2 of the Ornstein–Uhlenbeck semigroup R_t defined by (10.30).

Proposition 10.34 *We have*

$$D(L_2) = \left\{ \varphi \in W^{2,2}(H, \mu) : \int_H |(-A)^{1/2} D\varphi|^2 d\mu < \infty \right\}. \qquad (10.54)$$

Proof. We have

$$(-A)^{1/2} D\varphi(x) = \frac{1}{\sqrt{2}} \sum_{k=1}^{\infty} \frac{1}{\sqrt{\lambda_k}} D_k\varphi(x)e_k.$$

By (10.19) it follows that

$$\int_H |(-A)^{1/2} D\varphi|^2 d\mu = \frac{1}{2} \sum_{\gamma \in \Gamma} \langle \gamma, \lambda^{-2} \rangle |\varphi_\gamma|^2.$$

Thus it is enough to prove that if $\varphi \in D(L)$ the following identity holds

$$\sum_{\gamma \in \Gamma} |\langle \gamma, \lambda^{-1} \rangle|^2 |\varphi_\gamma|^2 = \sum_{h,k=1}^{\infty} \int_H |D_h D_k\varphi|^2 d\mu + \sum_{\gamma \in \Gamma} \langle \gamma, \lambda^{-2} \rangle |\varphi_\gamma|^2. \tag{10.55}$$

We have in fact for any $h \in \mathbb{N}$,

$$D_h^2 H_\gamma = \frac{\sqrt{\gamma_h(\gamma_h - 1)}}{\lambda_h} H_{\gamma^{(h,h)}},$$

where

$$\gamma_n^{(h,h)} = \begin{cases} \gamma_n & \text{if } n \neq h, \\ \gamma_h - 2 & \text{if } n = h, \end{cases}$$

and we have set $H_{-2}(\xi) = 0$. Moreover if $h, k \in \mathbb{N}$ with $h \neq k$ we have

$$D_h D_k H_\gamma = \sqrt{\frac{\gamma_h \gamma_k}{\lambda_h \lambda_k}} H_{\gamma^{(h,k)}},$$

where

$$\gamma_n^{(h,k)} = \begin{cases} \gamma_n & \text{if } n \neq h, n \neq k \\ \gamma_h - 1 & \text{if } n = h, \\ \gamma_k - 1 & \text{if } n = k. \end{cases}$$

It follows that

$$D_h^2 \varphi = \sum_{\gamma \in \Gamma} \frac{\sqrt{\gamma_h(\gamma_h - 1)}}{\lambda_h} \, \varphi_\gamma H_{\gamma^{(h,h)}},$$

and

$$D_h D_k H_\gamma \varphi = \sum_{\gamma \in \Gamma} \sqrt{\frac{\gamma_h \gamma_k}{\lambda_h \lambda_k}} \, \varphi_\gamma H_{\gamma^{(h,k)}}.$$

Therefore

$$\sum_{h=1}^{\infty} \int_H |D_h \varphi|^2 d\mu = \int_{\gamma \in \Gamma} \frac{\gamma_h(\gamma_h - 1)}{\lambda_h^2} |\varphi_\gamma|^2, \qquad (10.56)$$

and

$$\sum_{h,k=1}^{\infty} \int_H |D_h D_k \varphi|^2 d\mu = \sum_{\gamma \in \Gamma} [\langle \gamma, \lambda^{-1} \rangle]^2 |\varphi_\gamma|^2 - \sum_{\gamma \in \Gamma} \langle \gamma, \lambda^{-2} \rangle |\varphi_\gamma|^2. \quad (10.57)$$

Now (10.55) follows from (10.56) and (10.57). □

11

Gradient systems

11.1 Introduction and setting of the problem

We are given a separable real Hilbert space H (norm $|\cdot|$, inner product $\langle \cdot, \cdot \rangle$), a self-adjoint strictly negative operator $A : D(A) \subset H \to H$ such that A^{-1} is of trace class and a non-negative mapping $U : H \to [0, +\infty]$ (the *potential*).

We consider the Gaussian measure $\mu = N_Q$ where $Q = -\frac{1}{2} A^{-1}$ and the probability measure (*Gibbs measure*)

$$\nu(dx) = Z^{-1} e^{-2U(x)} \mu(dx), \quad x \in H,$$

where Z is the normalization constant

$$Z = \int_H e^{-2U(x)} \mu(dx).$$

Then we consider the following Kolmogorov operator

$$N_0 \varphi = L_2 \varphi - \langle D_x U, D_x \varphi \rangle, \quad \varphi \in \mathscr{E}_A(H),$$

where L_2 is the Ornstein–Uhlenbeck operator

$$L_2 \varphi = \frac{1}{2} \operatorname{Tr} [D_x^2 \varphi] + \langle x, A D_x \varphi \rangle, \quad \varphi \in \mathscr{E}_A(H),$$

studied in section 8.3. [1]

The goal of the chapter is to show that, under suitable assumptions on U, N_0 is essentially self-adjoint in $L^2(H, \nu)$. Then, denoting by N_2 the closure of N_0, we study several properties of the Markov semigroup $P_t = e^{tN_2}$.

---

[1] $\mathscr{E}_A(H)$ is the space of exponential functions defined in (8.28).

We will not attempt to develop a general theory of Gibbs measures; we will discuss only a significant situation, see section 11.2 for a motivating example. For an introduction to spin systems and Gibbs measures see for instance [25] and [10, Chapter 11].

11.1.1 Assumptions and notations

Concerning A we shall assume that

Hypothesis 11.1
(i) A is self-adjoint and there exists $\omega > 0$ such that

$$\langle Ax, x \rangle \leq -\omega |x|^2, \quad x \in D(A).$$

(ii) There exists $\beta \in (0,1)$ such that $\mathrm{Tr}[(-A)^{\beta-1}] < +\infty$.

We set $Q = -\frac{1}{2} A^{-1}$ and consider the nondegenerate Gaussian measure $\mu = N_Q$. We denote by (e_k) a complete orthonormal system of eigenvectors of Q and by (λ_k) the corresponding eigenvalues. For any $x \in H$ we set $x_k = \langle x, e_k \rangle$, $k \in \mathbb{N}$.

Moreover, we consider the Ornstein–Uhlenbeck semigroup R_t in $L^p(H, \mu)$ where $p \geq 1$ (introduced in Chapters 8 and 10, see in particular (10.30)),

$$R_t \varphi(x) = \int_H \varphi(e^{tA}x + y) N_{Q_t}(dy), \quad x \in H, \ t \geq 0, \ \varphi \in B_b(H), \ (11.1)$$

where
$$Q_t = Q(1 - e^{2tA}), \quad t \geq 0. \tag{11.2}$$

We recall that, for any $p \geq 1$, R_t is a strongly continuous semigroup of contractions in $L^p(H, \mu)$ and that the space of all exponential functions $\mathscr{E}_A(H)$ is a core for its infinitesimal generator L_p (Theorem 8.22). Finally,

$$L_p \varphi = \frac{1}{2} \, \mathrm{Tr} \, [D_x^2 \varphi] + \langle x, A D_x \varphi \rangle, \quad \varphi \in \mathscr{E}_A(H). \tag{11.3}$$

Concerning the potential $U \colon H \to \mathbb{R} \cup \{+\infty\}$ we shall assume that

Hypothesis 11.2
(i) U belongs to $W^{1,2}(H, \mu)$ and it is non-negative.
(ii) Setting

$$Z := \int_H e^{-2U(y)} \mu(dy)$$

and

$$\rho(x) = Z^{-1} e^{-2U(x)}, \quad x \in H,$$

we have

$$\rho, \sqrt{\rho} \in W^{1,2}(H, \mu). \tag{11.4}$$

Then we define the probability measure ν on $(H, \mathscr{B}(H))$,

$$\nu(dx) = \rho(x)\mu(dx). \tag{11.5}$$

Let us explain the content of this chapter. After a motivating example presented in section 11.2, section 11.3 is devoted to the definition of the Sobolev space $W^{1,2}(H, \nu)$. In section 11.4 we introduce the linear operator

$$N_0\varphi = L_2\varphi - \langle D_xU, D_x\varphi \rangle, \quad \varphi \in \mathscr{E}_A(H), \tag{11.6}$$

and show that it is symmetric in $L^2(H, \nu)$ and consequently, closable (see Step 1 in the proof of Theorem A.21).

In order to prove that N_0 is essentially self-adjoint we need some other tools as the notion of cylindrical Wiener process and some results on stochastic differential equations in Hilbert spaces which we introduce in section 11.5.

Section 11.6 is devoted to proving, under the additional assumption that U is convex and lower semicontinuous, that N_0 is *essentially self-adjoint* (the reader interested to the case when U is not convex can look at [10, Chapter 10]).

In this section an important rôle will be played by the *Moreau–Yosida* approximations U_α, $\alpha > 0$, of U (we shall recall some properties of U_α at the end of this section) and by the approximating equation, [2]

$$\lambda\varphi_\alpha - L_2\varphi_\alpha - \langle D_xU_\alpha, D_x\varphi_\alpha \rangle = f, \quad \alpha > 0. \tag{11.7}$$

We shall show that the solution of equation (11.7) is given by the formula

$$\varphi_\alpha(x) = \int_0^{+\infty} e^{-\lambda t}\mathbb{E}[f(X_\alpha(t, x))]dt, \quad x \in H,$$

where $X_\alpha(t, x)$ is the solution to the stochastic differential equation

$$\begin{cases} dX_\alpha = (AX_\alpha - D_xU_\alpha(X_\alpha))dt + dW(t) \\ \\ X_\alpha(0) = x, \end{cases} \tag{11.8}$$

and $W(t)$ is a cylindrical Wiener process in H.

Finally, in section 11.7 we study the semigroup generated by the closure of N_0, the Poincaré and log-Sobolev inequalities and the compactness of the embedding of $W^{1,2}(H, \nu)$ into $L^2(H, \nu)$.

---

[2] Indeed, we shall need a further approximation, see section 11.6.

Remark 11.3 Once we have defined the space $W^{1,2}(H, \nu)$, we can consider the symmetric Dirichlet form

$$a(\varphi, \psi) = \frac{1}{2} \int_H \langle D_x \varphi, D_x \psi \rangle d\nu, \quad \varphi, \psi \in W^{1,2}(H, \nu). \qquad (11.9)$$

Since a is obviously coercive, by the Lax–Milgram theorem there exists a negative self-adjoint operator N_2 in $L^2(H, \nu)$ such that

$$a(\varphi, \psi) = -\frac{1}{2} \int_H N_2 \varphi \, \psi \, d\nu, \quad \varphi, \psi \in D(N_2). \qquad (11.10)$$

One can show that N_2 can be identified with the closure of the differential operator N_0 above, see e.g. [24] and references therein.

11.1.2 Moreau–Yosida approximations

Let $U \colon H \to (-\infty, +\infty]$ be convex and lower semi-continuous. Then the Moreau–Yosida approximations U_α of U are defined by,

$$U_\alpha(x) = \inf \left\{ U(y) + \frac{1}{2\alpha}|x - y|^2 : y \in H \right\}, \quad x \in H, \ \alpha > 0.$$

Let us recall some properties of U_α, for proofs see e.g. [3].

Lemma 11.4 *Let $\alpha > 0$. Then the following properties hold.*

(i) U_α *is convex, differentiable, and* $U_\alpha(x) \le U(x)$ *for all* $x \in H$.
(ii) *We have*
$$|D_x U_\alpha(x)| \le |D_x U(x)|, \quad \text{for all } x \in H,$$

where $|D_x U(x)|$ *is the element of the subdifferential* $\partial U(x)$ *(which is a non-empty convex closed set) of minimal norm.*
(iii) *We have*

$$\lim_{\alpha \to 0} U_\alpha(x) = U(x), \quad \text{for all } x \in H,$$

$$\lim_{\alpha \to 0} D_x U_\alpha(x) = D_x U(x), \quad \text{for all } x \in H.$$

11.2 A motivating example

Let us consider the linear operator in $H := L^2(0, 1)$ defined as

$$\begin{cases} Ax = D_\xi^2 x, & x \in D(A), \\ D(A) = H^2(0, 1) \cap H_0^1(0, 1), \end{cases} \qquad (11.11)$$

where $\xi \in [0, 1]$, $H^k(0, 1)$, $k = 1, 2$, are the usual Sobolev spaces and

$$H_0^1(0, 1) = \{x \in H^1(0, 1) : x(0) = x(1) = 0\}.$$

It is well known that A is self-adjoint and that

$$Ae_k = -\pi^2 k^2 e_k, \quad k \in \mathbb{N},$$

where (e_k) is the complete orthonormal system on H defined as

$$e_k(\xi) = \sqrt{\frac{2}{\pi}} \sin(k\pi\xi), \quad \xi \in [0, 1], \ k \in \mathbb{N}. \tag{11.12}$$

It is easy to see that A fulfills Hypothesis 11.1 with $\omega = \pi^2$ and with β equal to any number in $(0, 1)$. So, we have $Q = -\frac{1}{2} A^{-1}$, and $\lambda_k = \frac{1}{2\pi^2 k^2}$, $k \in \mathbb{N}$.

Remark 11.5 The measure $\mu = N_Q$ is the law of the *Brownian bridge* in $[0, 1]$, see subsection 3.4.1. Therefore μ is concentrated on the space,

$$\{x \in C([0, 1]) : x(0) = x(1) = 0\}$$

(we shall not use this fact in what follows).

We are interested in the following potential

$$U_m(x) = \begin{cases} \dfrac{1}{2m} \displaystyle\int_0^1 |x(\xi)|^{2m} d\xi, & \text{if } x \in L^{2m}(0, 1), \\ +\infty, & \text{if } x \notin L^{2m}(0, 1), \end{cases} \tag{11.13}$$

where $m \in \mathbb{N}$ is given. We are going to show that U fulfills Hypothesis 11.2. For this we need some preliminary result on random variables in $L^2(0, 1)$, interesting in itself.

Remark 11.6 The measure ν, defined by

$$\nu(dx) = Z^{-1} e^{-2U_m} \mu(dx),$$

is the *Gibbs measure* corresponding to the stochastic *reaction-diffusion* equation

$$dX = (AX - X^{2m-1})dt + dW(t),$$

where $W(t)$ is a cylindrical Wiener process in $L^2(0, 1)$, see subsection 11.5.1 below and [7].

11.2.1 Random variables in $L^2(0,1)$

We set here $H = L^2(0,1)$. Let us first define the Dirac delta at ξ for any $\xi \in [0,1]$ on $L^2(H,\mu)$. For any $x \in H$ and $N \in \mathbb{N}$ we set

$$\delta_\xi^N(x) = x_N(\xi) = \sum_{k=1}^N \langle x, e_k \rangle e_k(\xi), \quad \xi \in [0,1],$$

where the system (e_k) is defined by (11.12). Consequently,

$$\delta_\xi^N(x) = \left\langle x, \sum_{k=1}^N e_k(\xi)e_k \right\rangle = \frac{1}{\sqrt{2\pi}} \langle x, Q^{-1/2}\eta_{N,\xi} \rangle = W_{\eta_{N,\xi}}(x),$$

where

$$\eta_{N,\xi} = \frac{1}{\sqrt{2\pi}} \sum_{k=1}^N \frac{1}{k} e_k(\xi)e_k, \quad \xi \in [0,1].$$

Notice that $\eta_{N,\xi} \in W^{1,2}(0,1)$ for all $\xi \in [0,1]$ and that

$$\lim_{N\to\infty} \eta_{N,\xi} = \eta_\xi \quad \text{in } H, \quad \text{for all } \xi \in [0,1],$$

where

$$\eta_\xi = \frac{1}{\sqrt{2\pi}} \sum_{k=1}^\infty \frac{1}{k} e_k(\xi)e_k, \quad \xi \in [0,1]. \tag{11.14}$$

It follows that there exists the limit

$$\lim_{N\to\infty} W_{\eta_{N,\xi}} = W_{\eta_\xi} \quad \text{in } L^2(H,\mu).$$

So, we define

$$\delta_\xi(x) = x(\xi) := W_{\eta_\xi}(x), \quad \text{for all } \xi \in [0,1] \text{ and for } \mu\text{-almost } x \in H.$$

Notice that, by the Parseval identity, we have

$$|\eta_\xi|^2 = \frac{1}{2\pi^2} \sum_{k=1}^\infty \frac{1}{k^2} |e_k(\xi)|^2 \le \frac{1}{6\pi}, \quad \xi \in [0,1]. \tag{11.15}$$

Exercise 11.7 Show that η_ξ is continuous in ξ.

Now, we shall write the potential U_m as

$$U_m(x) = \frac{1}{2m} \int_0^1 W_{\eta_\xi}^{2m}(x)d\xi, \quad x \in H.$$

Proposition 11.8 U_m belongs to $L^p(H, \mu)$ for any $p \geq 1$.

Proof. Let us consider the case $p = 1$. Then, by the Fubini theorem and taking into account (9.11) and (11.15), we find that

$$\int_H \left(\int_0^1 W_{\eta_\xi}^{2m}(x)d\xi \right) \mu(dx) = \int_0^1 \left(\int_H W_{\eta_\xi}^{2m}(x)\mu(dx) \right) d\xi$$

$$\leq C_m \int_0^1 (1 + |\eta_\xi|^m)d\xi < +\infty.$$

So, $U_m \in L^1(H, \mu)$. The case $p > 1$ can be handled similarly. $\square$

Proposition 11.9 U_m belongs to $W^{1,2}(H, \mu)$ and [3]

$$D_x U_m(x)(\xi) = W_{\eta_{N,\xi}}^{2m-1}(x) = x^{2m-1}(\xi), \quad \mu\text{-a.e.} \tag{11.16}$$

Proof. Set

$$U_{m,N}(x) = \frac{1}{2m} \int_0^1 W_{\eta_{N,\xi}}^{2m}(x)d\xi, \quad x \in H.$$

Then $U_{m,N}$ is a C^1 function and, for any $h \in H$, we have

$$D_x U_{m,N}(x) \cdot h = \int_0^1 W_{\eta_{N,\xi}}^{2m-1}(x)h(x)d\xi = \int_0^1 x_N^{2m-1}(\xi)h(\xi)d\xi.$$

It is easy to see that

$$\lim_{N \to \infty} U_{m,N} = U_m \quad \text{in } L^2(H, \mu),$$

$$\lim_{N \to \infty} D_x U_{m,N} = D_x U_m \quad \text{in } L^2(H, \mu; H).$$

Consequently, U_m belongs to $W^{1,2}(H, \mu)$ and (11.16) holds. $\square$

Exercise 11.10 Prove that U_m belongs to $W^{1,p}(H, \mu)$ for all $p \geq 1$.

We can now check that Hypothesis 11.2 is fulfilled. We first notice that $\rho = Z^{-1}e^{-2U_m}$ is bounded. Moreover, we prove the result

Proposition 11.11 ρ and $\sqrt{\rho}$ belong to $W^{1,2}(H, \mu)$ and

$$D_x \rho = -2D_x U_m \rho. \tag{11.17}$$

Proof. Since $\rho = Z^{-1}e^{-2U_m}$ and $U_m \in W^{1,2}(H, \mu)$, we can apply Proposition 10.8 to conclude that $\rho \in W^{1,2}(H, \mu)$ and (11.17) holds. In a similar way, one can prove that $\sqrt{\rho} \in W^{1,2}(H, \mu)$. $\square$

---

[3] Throughout this chapter we shall write $\overline{D} = D = D_x$ for simplicity.

11.3 The Sobolev space $W^{1,2}(H, \nu)$

We proceed here to define the Sobolev space $W^{1,2}(H, \mu)$ as in Chapter 10. So, we start by proving an integration by parts formula.

Lemma 11.12 *Assume that Hypotheses* 11.1 *and* 11.2 *hold. Let* $\varphi, \psi \in \mathscr{E}(H)$ *and* $k \in \mathbb{N}$. *Then we have*

$$\int_H D_k \varphi \, \psi \, d\nu = -\int_H \varphi \, D_k \psi \, d\nu + \int_H \left[\frac{x_k}{\lambda_k} - D_k \log \rho \right] \varphi \, \psi \, d\nu. \quad (11.18)$$

Proof. Write

$$\int_H D_k \varphi \, \psi \, d\nu = \int_H D_k \varphi \, \psi \, \rho \, d\mu.$$

Since $\psi \rho \in W^{1,2}(H, \mu)$, we can apply the integration by parts formula (10.1), and we get

$$\int_H D_k \varphi \, \psi \, d\nu = -\int_H \varphi D_k(\psi \rho) d\mu + \frac{1}{\lambda_k} \int_H x_k \varphi \, \psi \, d\nu$$

$$= -\int_H \varphi \, D_k \psi \, d\nu - \int_H \varphi \, \psi \, D_k \log \rho \, d\nu + \frac{1}{\lambda_k} \int_H x_k \varphi \, \psi \, d\nu,$$

which yields (11.18). $\square$

Before proving closability of the derivative, we need a lemma.

Lemma 11.13 *Assume that Hypotheses* 11.1 *and* 11.2 *hold. Then we have* $D_x U = -\frac{1}{2} D_x \log \rho \in L^2(H, \nu; H)$.

Proof. Since

$$U = -\frac{1}{2} \log \rho - \frac{1}{2} \log Z,$$

and $D_x \log \rho \in L^2(H, \mu; H)$, we have

$$D_x U = -\frac{1}{2} D_x \log \rho = -\frac{1}{2} \frac{D_x \rho}{\rho}.$$

Consequently,

$$\int_H |D_x \log \rho|^2 d\nu = \int_H \frac{|D_x \rho|^2}{\rho} d\mu = 4 \int_H |D_x \sqrt{\rho}|^2 d\mu.$$

$\square$

Proposition 11.14 *Assume that Hypotheses* 11.1 *and* 11.2 *hold. Then for any* $k \in \mathbb{N}$ *the operator* D_k *is closable on* $L^2(H, \nu)$.

We shall still denote by D_k its closure.

Proof. Let (φ_n) be a sequence in $\mathscr{E}(H)$ and $g \in L^2(H, \nu)$ such that

$$\varphi_n \to 0, \quad D_k\varphi_n \to g \quad \text{in } L^2(H, \nu).$$

We have to show that $g = 0$. Let $\psi \in \mathscr{E}(H)$. Then by Lemma 11.12 we have

$$\int_H D_k\varphi_n \, \psi \, d\nu = -\int_H \varphi_n \, D_k\psi \, d\nu + \int_H \frac{x_k}{\lambda_k} \varphi_n \, \psi \, d\nu$$

$$- \int_H D_k \log \rho \, \varphi_n \, \psi \, d\nu. \tag{11.19}$$

Moreover, taking into account Lemma 11.13, we have

$$\begin{cases} \lim\limits_{n \to \infty} \int_H D_k\varphi_n \, \psi \, d\nu = \int_H g\psi d\nu, \\[2mm] \lim\limits_{n \to \infty} \int_H \varphi_n \, D_k\psi \, d\nu = 0 \\[2mm] \lim\limits_{n \to \infty} \int_H D_k \log \rho \, \varphi_n \, \psi \, d\nu = 0. \end{cases} \tag{11.20}$$

We also claim that

$$\lim\limits_{n \to \infty} \int_H x_k\varphi_n\psi d\nu = 0. \tag{11.21}$$

We have in fact

$$\left| \int_H x_k\varphi_n\psi d\nu \right|^2 \le \int_H \varphi_n^2 d\nu \int_H x_k^2\psi^2 \rho d\mu$$

$$\le \int_H \varphi_n^2 d\nu \left(\int_H \rho^2 d\mu \right)^{1/2} \left(\int_H x_k^4\psi^4 d\mu \right)^{1/2} \to 0,$$

as $n \to \infty$. So, (11.21) follows. Now (11.20) and (11.21) yield

$$\int_H g\psi d\nu = 0,$$

so that the conclusion follows from the arbitrariness of g. $\square$

We now define $W^{1,2}(H, \nu)$ as the linear space of all functions $\varphi \in L^2(H, \nu)$ such that $D_k\varphi \in L^2(H, \nu)$ for all $k \in \mathbb{N}$ and

$$\sum_{k=1}^{\infty} \int_H |D_k\varphi(x)|^2 \nu(dx) < +\infty.$$

$W^{1,2}(H, \nu)$, endowed with the inner product,

$$\langle \varphi, \psi \rangle_{W^{1,2}(H,\nu)} = \langle \varphi, \psi \rangle_{L^2(H,\nu)} + \sum_{k=1}^{\infty} \int_H D_k \varphi \, D_k \psi \, d\nu,$$

is a Hilbert space.

If $\varphi \in W^{1,2}(H, \nu)$ we set

$$D_x \varphi(x) = \sum_{k=1}^{\infty} D_k \varphi(x) e_k, \quad \nu\text{-a.e. in } H.$$

Since

$$|D_x \varphi(x)|^2 = \sum_{k=1}^{\infty} |D_k \varphi(x)|^2, \quad \nu\text{-a.e. in } H,$$

the series is convergent for ν-almost all $x \in H$. We call $D_x \varphi(x)$ the *gradient* of φ at x.

In an analogous way we can define $W^{k,2}(H, \nu)$, $k \geq 2$.

11.4 Symmetry of the operator N_0

Let us consider the linear operator

$$N_0 \varphi = L_2 \varphi - \langle D_x U, D_x \varphi \rangle, \quad \varphi \in \mathscr{E}_A(H). \tag{11.22}$$

We want to consider N_0 as an operator in $L^2(H, \nu)$. This is possible thanks to the following lemma.

Lemma 11.15 *Assume that Hypotheses 11.1 and 11.2 hold. Then*

(i) N_0 respects ν-classes. That is if φ and ψ are two functions in $\mathscr{E}_A(H)$ such that $\varphi = \psi$ ν-a.e. then φ is identically equal to ψ.

(ii) $N_0 \varphi \in L^2(H, \nu)$ for all $\varphi \in \mathscr{E}_A(H)$.

Proof. Let us prove (i). Let $\varphi, \psi \in \mathscr{E}_A(H)$ such that $\varphi = \psi$, ν-a.e. Then we have

$$\varphi(x)\rho(x) = \psi(x)\rho(x), \quad \mu\text{-a.e.}$$

On the other hand, $\rho(x) = Z^{-1} e^{-2U(x)} > 0$ μ-a.e. since $U(x) < +\infty$ μ-a.e. thanks to Hypothesis 11.2(i). So, we have $\varphi = \psi$ μ-a.e. Since the measure μ is full this implies that φ is identically equal to ψ.

Let us finally prove (ii). Write

$$\int_H |N_0 \varphi|^2 d\nu \leq 2 \int_H |L_2 \varphi|^2 d\nu + 2 \int_H |\langle D_x U, D_x \varphi \rangle|^2 d\nu.$$

Let $a, b > 0$ be such that

$$|L_2\varphi(x)| \leq a + b|x|, \quad x \in H,$$

then we have

$$\int_H |L_2\varphi|^2 d\nu \leq 2a^2 + 2b^2 \int_H |x|^2 \rho d\mu$$

$$\leq 2a^2 + 2b^2 \left(\int_H \rho^2 d\mu\right)^{1/2} \left(\int_H |x|^4 d\mu\right)^{1/2} < +\infty.$$

Moreover,

$$\int_H |\langle D_x U, D_x\varphi\rangle|^2 d\nu \leq \|\varphi\|_1 \int_H |D_x U|^2 d\nu < +\infty.$$

So, the conclusion follows. $\square$

Proposition 11.16 *Assume that Hypotheses 11.1 and 11.2 hold. Then for all $\varphi, \psi \in \mathscr{E}_A(H)$ we have*

$$\int_H N_0\varphi \, \psi d\nu = -\frac{1}{2} \int_H \langle D_x\varphi, D_x\psi\rangle d\nu. \tag{11.23}$$

Therefore N_0 is symmetric in $L^2(H, \nu)$.

Proof. Let $\varphi, \psi \in \mathscr{E}_A(H)$. Then we have

$$\int_H N_0\varphi \, \psi \, d\nu = \int_H L_2\varphi \, \psi \, \rho d\mu - \int_H \langle D_x U, D_x\varphi\rangle \psi d\nu.$$

Since $\rho \in W^{1,2}(H, \mu)$ we have $\psi\rho \in W^{1,2}(H, \mu)$ and

$$\int_H L\varphi\psi\rho d\mu = -\frac{1}{2} \int_H \langle D_x\varphi, D_x(\psi\rho)\rangle d\mu$$

$$= -\frac{1}{2} \int_H \langle D_x\varphi, D_x\psi\rangle d\nu - \frac{1}{2} \int_H \langle D_x\varphi, D_x \log\rho\rangle \psi d\nu$$

$$= -\frac{1}{2} \int_H \langle D_x\varphi, D_x\psi\rangle d\nu + \frac{1}{2} \int_H \langle D_x\varphi, D_x U\rangle \psi d\nu,$$

and the conclusion follows. $\square$

Since N_0 is dissipative in $L^2(H, \nu)$, it is closable (see Theorem A.21); we shall denote by N_2 the closure of N_0.

11.5 Some complements on stochastic differential equations

11.5.1 Cylindrical Wiener process and stochastic convolution

Let us define a cylindrical Wiener process on the probability space $(\Omega, \mathscr{F}, \mathbb{P})$ where $\Omega = \mathscr{H} = L^2(0, +\infty; H)$, $\mathscr{F} = \mathscr{B}(\mathscr{H})$, and $\mathbb{P} = N_{\mathscr{Q}}$, where $\mathscr{Q}$ is any operator in $L_1^+(\mathscr{H})$ such that $\mathrm{Ker}\, \mathscr{Q} = \{0\}$. Then we set [4]

$$\beta_k(t) = W_{\mathbf{1}_{[0,t]}e_k}, \quad t \geq 0, \ k \in \mathbb{N}.$$

Clearly (β_k) is a sequence of independent standard Brownian motions in $(\Omega, \mathscr{F}, \mathbb{P})$.

Now the cylindrical Wiener process W can be defined formally by setting

$$W(t) = \sum_{k=1}^{\infty} e_k \beta_k(t), \quad t \geq 0.$$

Note that this series is $\mathbb{P}$-a.s. divergent since

$$\mathbb{E}(|W(t)|^2) = \sum_{k=1}^{\infty} t = +\infty, \quad t \geq 0.$$

However, the *stochastic convolution*

$$W_A(t) = \int_0^t e^{(t-s)A} dW(s) = \sum_{k=1}^{\infty} \int_0^t e^{(t-s)A} e_k d\beta_k(s), \tag{11.24}$$

is well defined as the next proposition shows.

Proposition 11.17 *Assume that Hypothesis 11.1 holds. Then for each $t > 0$ the series defined by (11.24) is convergent on $L^2(\Omega, \mathscr{F}, \mathbb{P}; H)$ to a Gaussian random variable $W_A(t)$ with mean 0 and covariance operator Q_t, where*

$$Q_t = Q(1 - e^{2tA}), \quad t > 0.$$

We shall denote by μ_t the Gaussian measure N_{Q_t}.

Proof. Let $t > 0$, and $n, p \in \mathbb{N}$. Taking into account the independence of the real Brownian motions (β_k), we find that

$$\mathbb{E}\left| \sum_{k=n+1}^{n+p} \int_0^t e^{(t-s)A} e_k d\beta_k(s) \right|^2 = \sum_{k=n+1}^{n+p} \int_0^t |e^{(t-s)A} e_k|^2 ds.$$

---

[4] See Chapter 1 for the definition of the white noise mapping W.

Since

$$\sum_{k=1}^{\infty} \int_0^t |e^{(t-s)A}e_k|^2 ds = \operatorname{Tr} Q_t < +\infty,$$

the series

$$\sum_{k=1}^{\infty} \int_0^t e^{(t-s)A}e_k d\beta_k(s),$$

is convergent in $L^2(\Omega, \mathscr{F}, \mathbb{P}; H)$, to a random variable $W_A(t) \in L^2(\Omega, \mathscr{F}, \mathbb{P}; H)$.

By Proposition 1.16 it follows that $W_A(t)$, as the limit of Gaussian random variables, is a Gaussian random variable as well. By a simple computation we find finally

$$\mathbb{E}\left[\langle W_A(t), h\rangle \langle W_A(t), k\rangle\right] = \langle Q_t h, k\rangle, \quad h, k \in H.$$

Therefore the covariance operator of $W_A(t)$ is Q_t and consequently the law of $W_A(t)$ is N_{Q_t}. □

Now we are going to show that, thanks to Hypothesis 11.1(ii), $W_A(\cdot)$ is continuous $\mathbb{P}$-almost surely.

We shall use again the factorization method introduced in section 3.1. Using identity (3.1) we can write the stochastic convolution W_A as follows,

$$W_A(t) = \frac{\sin \pi \alpha}{\pi} \int_0^t e^{(t-\sigma)A}(t-\sigma)^{\alpha-1}Y(\sigma)d\sigma, \tag{11.25}$$

where $\alpha \in (0,1)$ and

$$Y(\sigma) = \int_0^\sigma e^{(\sigma-s)A}(\sigma-s)^{-\alpha}dW(s), \quad \sigma \geq 0. \tag{11.26}$$

To prove continuity of W_A we need an elementary analytic lemma which is a generalization of Lemma 3.2.

Lemma 11.18 *Let* $m > 1, T > 0, \alpha \in (1/(2m), 1)$ *and* $g \in L^{2m}(0, T; H)$. *Set*

$$G(t) = \int_0^t e^{(t-\sigma)A}(t-\sigma)^{\alpha-1}g(\sigma)d\sigma, \quad t \in [0, T].$$

Then $G \in C([0, T]; H)$ *and there is* $C_{m,T} > 0$ *such that*

$$|G(t)| \leq C_{m,T}\|g\|_{L^{2m}(0,T;H)}, \quad t \in [0, T]. \tag{11.27}$$

Proof. Let $t \in [0, T]$, then by Hölder's inequality we have (notice that $2m\alpha - 1 > 0$),

$$|G(t)| \leq M_T \left(\int_0^t (t-\sigma)^{(\alpha-1)\frac{2m}{2m-1}} d\sigma \right)^{\frac{2m-1}{2m}} |g|_{L^{2m}(0,T;H)}, \qquad (11.28)$$

which yields (11.27).

It remains to show the continuity of F. Continuity at 0 follows from (11.28). Let $t_0 > 0$. We are going to prove that F is continuous on $[\frac{t_0}{2}, T]$. Let us set for $\varepsilon < \frac{t_0}{2}$

$$G_\varepsilon(t) = \int_0^{t-\varepsilon} e^{(t-\sigma)A}(t-\sigma)^{\alpha-1} g(\sigma) d\sigma, \quad t \in [0, T].$$

G_ε is obviously continuous on $[\varepsilon, T]$. Moreover, using once again Hölder's inequality, we find

$$|G(t) - G_\varepsilon(t)| \leq \left(\frac{2m-1}{2m\alpha - 1} \right)^{\frac{2m-1}{2m}} \varepsilon^{\alpha - \frac{1}{2m}} |f|_{L^{2m}(0,T;H)}, \quad t \geq 0.$$

Thus $\lim_{\varepsilon \to 0} G_\varepsilon(t) = G(t)$, uniformly on $[\frac{t_0}{2}, T]$, and G is continuous as required. $\square$

Proposition 11.19 *Assume that Hypothesis* 11.1 *holds. Let* $m > 1$ *and* $T \in (0, +\infty)$. *Then there exists a constant* $C_{m,T} > 0$ *such that*

$$\mathbb{E} \left(\sup_{t \in [0,T]} |W_A(t)|^{2m} \right) \leq C_{m,T}. \qquad (11.29)$$

Moreover $W_A(\cdot)(\omega)$ *is* $\mathbb{P}$-*a.e. continuous on* $[0, T]$.

Proof. Let Y be defined by (11.26). Then, in view of Proposition 11.17, $Y(\sigma)$ is a Gaussian random variable N_{S_σ} for all $\sigma \in (0, T]$, where

$$S_\sigma x = \int_0^\sigma s^{-2\alpha} e^{2sA} x \, ds, \quad x \in H.$$

We set $\mathrm{Tr} \, S_\sigma = C_{\alpha,\sigma}$. Consequently, for any $m > 1$ there exists a constant $D_{m,\alpha} > 0$ such that

$$\mathbb{E} \left(|Y(\sigma)|^{2m} \right) \leq D_{m,\alpha} \sigma^m, \quad \sigma \in [0, T]. \qquad (11.30)$$

This implies that

$$\int_0^T \mathbb{E} \left(|Y(\sigma)|^{2m} \right) d\sigma \leq \frac{D_{m,\alpha}}{m+1} T^{m+1},$$

so that $Y(\cdot)(\omega) \in L^{2m}(0, T; H)$ for almost all $\omega \in \Omega$. Therefore $W_A(\cdot)(\omega) \in C([0, T]; H)$ for almost all $\omega \in \Omega$. Moreover, recalling (11.28), we have

$$\sup_{t\in[0,T]} |W_A(t)|^{2m} \leq \left(\frac{M_T}{\pi}\right)^{2m} \left(\frac{2m-1}{2m\alpha-1}\right)^{2m-1} T^{2m\alpha-1} \int_0^T |Y(\sigma)|^{2m} d\sigma.$$

Now (11.29) follows, taking expectation. □

11.5.2 Stochastic differential equations

Consider the following stochastic differential equation

$$\begin{cases} dX(t) = (AX(t) + F(X(t)))dt + dW(t), \\ \\ X(0) = x \in H, \end{cases} \tag{11.31}$$

where $F: H \to H$ is Lipschitz continuous. We shall denote by K the Lipschitz constant of F, so that

$$|F(x) - F(y)| \leq K|x - y|, \quad x, y \in H.$$

By a *mild* solution of problem (11.31) on $[0, T]$ we mean a stochastic process $X = C([0, T]; L^2(\Omega, \mathscr{F}, \mathbb{P}; H))$ such that [5]

$$X(t) = e^{tA}x + \int_0^t e^{(t-s)A} F(X(s))ds + W_A(t), \tag{11.32}$$

where W_A is the stochastic convolution defined before.

By $C([0, T]; L^2(\Omega, \mathscr{F}, \mathbb{P}; H))$ we mean the space of all stochastic processes $X(t)$ such that,

(i) $X(t) \in L^2(\Omega, \mathscr{F}, \mathbb{P}; H)$ for all $t \in [0, T]$.
(ii) The mapping $[0, T] \to L^2(\Omega, \mathscr{F}, \mathbb{P}; H)$, $t \mapsto X(t)$, is continuous.

$C([0, T]; L^2(\Omega, \mathscr{F}, \mathbb{P}; H))$, endowed with the norm

$$\|X\|_{C([0,T];L^2(\Omega,\mathscr{F},\mathbb{P};H))} := \sup_{t\in[0,T]} \left(\mathbb{E}(|X(t)|^2)\right)^{1/2},$$

is a Banach space.

Proposition 11.20 *For any $x \in H$ equation (11.32) has a unique mild solution X.*

---

[5] We do not need here that the process is adapted to W.

Proof. Setting $Z_T = C([0,T]; L^2(\Omega, \mathscr{F}, \mathbb{P}; H))$, (11.32) is equivalent to the following equation in Z_T,

$$X(t) = e^{tA}x + \gamma(X)(t) + W_A(t), \quad t \in [0,T],$$

where

$$\gamma(X)(t) = \int_0^t e^{(t-s)A}F(X(s))ds, \quad X \in Z_T, \quad t \in [0,T].$$

It is easy to check that γ maps Z_T into itself and that

$$\|\gamma(X) - \gamma(X_1)\|_{Z_T} \leq T\|X - X_1\|_{Z_T}, \quad X, X_1 \in Z_T.$$

Now let $T_1 \in (0,T]$ be such that $T_1 < 1$. Then γ is a contraction on Z_{T_1}. Therefore, equation (11.32) has a unique mild solution on $[0,T_1]$. By a similar argument, one can show existence and uniqueness on $[T_1, 2T_1]$ and so on. $\square$

We shall denote by $X(\cdot, x)$ the mild solution to problem (11.32). Let us study the continuous dependence of $X(\cdot, x)$ by x.

Proposition 11.21 Let $x, y \in H$. Then for any $x, y \in H$ we have

$$|X(t,x) - X(t,y)| \leq e^{TK}|x - y|. \tag{11.33}$$

Proof. In fact, for $t \in [0,T]$, we have

$$X(t,x) - X(t,y) = e^{tA}(x-y) + \int_0^t e^{(t-s)A}[F(X(s,x)) - F(X(s,y))]ds.$$

Consequently

$$|X(t,x) - X(t,y)| \leq |x-y| + K\int_0^t |X(s,x) - X(s,y)|ds$$

and (11.33) follows from Gronwall's lemma. $\square$

Now assume that F is of class C^2. Our goal is to prove that in this case the mild solution $X(t,x)$ of (11.32) is differentiable with respect to x and that for any $h \in H$ we have

$$D_x X(t,x) \cdot h = \eta^h(t,x), \quad t \geq 0, \ x \in H,$$

where $\eta^h(t,x)$ is the mild solution of the equation

$$\begin{cases} \dfrac{d}{dt}\,\eta^h(t,x) = A\eta^h(t,x) + D_x F(X(t,x)) \cdot \eta^h(t,x), \\ \eta^h(0,x) \quad = h. \end{cases} \tag{11.34}$$

This means that $\eta^h(t, x)$ is the solution of the integral equation

$$\eta^h(t, x) = e^{tA}h + \int_0^t e^{(t-s)A} D_x F(X(s, x)) \cdot \eta^h(s, x) ds, \quad t \geq 0.$$

It is easy to see, proceeding as for the proof of Proposition 11.20, that (11.34) has indeed a unique mild solution $\eta^h(t, x)$.

Theorem 11.22 *Assume, besides Hypothesis 11.1, that F is of class C^2 with bounded first and second derivatives.*[6] *Then $X(t, x)$ is differentiable with respect to x (P-a.s.) and for any $h \in H$ we have*

$$D_x X(t, x) \cdot h = \eta^h(t, x), \quad \text{P-a.s.} \tag{11.35}$$

where $\eta^h(t, x)$ is the mild solution of (11.34). Moreover,

$$|\eta^h(t, x)| \leq e^{(K-\omega)t}|h|, \quad t \geq 0, \tag{11.36}$$

where K is the Lipschitz constant of F.

Proof. Multiplying both sides of (11.34) by $\eta^h(t, x)$ and integrating over H yields

$$\frac{1}{2} \frac{d}{dt} |\eta^h(t, x)|^2 = \langle A\eta^h(t, x), \eta^h(t, x) \rangle + \langle D_x F(X(t, x)) \cdot \eta^h(t, x), \eta^h(t, x) \rangle$$

$$\leq (-\omega + K)|\eta^h(t, x)|^2,$$

which implies (11.36) from the Gronwall lemma.

Let us prove now that $\eta^h(t, x)$ fulfills (11.35). For this fix $T > 0$, $x \in H$ and $h \in H$ such that $|h| \leq 1$. We are going to show that there is $C_T > 0$ such that

$$|X(t + h, x) - X(t, x) - \eta^h(t, x)| \leq C_T |h|^2, \quad \text{P-a.s.} \tag{11.37}$$

Setting
$$r^h(t, x) = X(t, x + h) - X(t, x) - \eta^h(t, x),$$

we see that $r^h(t, x)$ satisfies the equation

$$r^h(t, x) = \int_0^t e^{(t-s)A}[F(X(s, x + h)) - F(X(s, x))] ds$$

$$- \int_0^t e^{(t-s)A} D_x F(X(s, x)) \cdot \eta^h(s, x) ds.$$

---

[6] In particular, F is Lipschitz continuous.

Consequently we have

$$r^h(t, x) = \int_0^t e^{(t-s)A} \left[\int_0^1 D_x F(\rho(\xi, s)) d\xi \right] \cdot (r^h(s, x) + \eta^h(s, x)) ds$$
$$- \int_0^t e^{(t-s)A} D_x F(X(s, x)) \cdot \eta^h(s, x) ds,$$

where

$$\rho(\xi, s) = \xi X(s, x + h) + (1 - \xi) X(s, x).$$

Therefore

$$r^h(t, x) = \int_0^t e^{(t-s)A} \left[\int_0^1 D_x F(\rho(\xi, s)) d\xi \right] \cdot r^h(s, x) ds$$
$$+ \int_0^t e^{(t-s)A} \left[\int_0^1 (D_x F(\rho(\xi, s))) \right.$$

$$\left. - D_x F(X(s, x))) d\xi \right] \cdot \eta^h(s, x) ds.$$

Setting $\gamma_T = \sup_{t \in [0,T]} e^{(-\omega + K)t}$, and taking into account (11.36), we find that

$$\left| \int_0^1 (D_x F(\rho(\xi, s)) - D_x F(X(s, x))) d\xi \right|$$
$$\leq \frac{1}{2} \|D^2 F\|_0 |h| |X(s, x + h) - X(s, x)|$$
$$\leq \frac{1}{2} \|D^2 F\|_0 |h| (|r^h(s, x)| + \gamma_T |h|).$$

It follows that

$$|r^h(t, x)| \leq (K + \|D^2 F\|_0) \int_0^t |r^h(s, x)| ds + T \gamma_T \|D^2 F\|_0 |h|^2.$$

Thus (11.37) follows from the Gronwall lemma. $\square$

11.6 Self-adjointness of N_2

In this section we assume that Hypotheses 11.1 and 11.2 are fulfilled and, in addition, that U is convex, nonnegative and lower semicontinuous. We shall approximate U by its Moreau–Yosida approximations U_α. However, U_α is not regular enough for our purposes (it is only C^1 with Lipschitz continuous derivative). So, we introduce a further approximation. We set

$$U_{\alpha,\beta}(x) := R_\beta U_\alpha(x) = \int_H U_\alpha(e^{\beta A} x + y) \mu_\beta(dy), \quad \alpha, \beta > 0, \quad (11.38)$$

where R_β is the Ornstein–Uhlenbeck semigroup defined by (11.1). By Remark 8.15, R_t fulfills (8.18) so that, in view of Theorem 8.16, $U_{\alpha,\beta}$ is C^∞ for any $\alpha, \beta > 0$. In particular, for any $h \in H$ we have

$$\langle D_x U_{\alpha,\beta}(x), h \rangle = \int_H \langle D_x U_\alpha(e^{\beta A} x + y), e^{\beta A} h \rangle \mu_t(dy), \quad \beta > 0. \quad (11.39)$$

It is easy to check that $D_x U_{\alpha,\beta}$ is Lipschitz continuous, of class C^∞ and that its derivatives of all orders are bounded in H. Moreover, in view of Lemma 11.4, we have

$$\lim_{\alpha \to 0} \lim_{\beta \to 0} D_x U_{\alpha,\beta}(x) = D_x U(x), \quad x \in H. \quad (11.40)$$

Our goal is to show that N_2 is self-adjoint or, equivalently, that for any $\lambda > 0$ and any $f \in L^2(H, \nu)$ there exists $\varphi \in D(N_2)$ solution of the equation

$$\lambda \varphi - N_2 \varphi = f. \quad (11.41)$$

We recall that, by a classical result due to Lumer and Phillips (see Theorem A.21 in Appendix A), in order to show that N_2 is self-adjoint it is enough to prove that the range of $\lambda - N_2$ is dense in $L^2(H, \nu)$ for one (and consequently for all) $\lambda > 0$.

It is convenient to introduce the approximating equation

$$\lambda \varphi_{\alpha,\beta} - L_2 \varphi_{\alpha,\beta} - \langle D_x U_{\alpha,\beta}, D_x \varphi_{\alpha,\beta} \rangle = f, \quad \alpha, \beta > 0. \quad (11.42)$$

To solve equation (11.42) we introduce the stochastic differential equation

$$\begin{cases} dX_{\alpha,\beta} = (AX_{\alpha,\beta} - D_x U_{\alpha,\beta}(X_{\alpha,\beta}))dt + dW(t) \\ \\ X_{\alpha,\beta}(0) = x, \end{cases} \quad (11.43)$$

where $W(t)$ is the cylindrical Wiener process in H introduced in section 11.5. Notice that equation (11.43) has a unique solution $X_{\alpha,\beta}$ in view of Proposition 11.20.

Now we are ready to prove the following result.

Proposition 11.23 *For any $f \in C_b^1(H)$, $\alpha, \beta > 0$, equation (11.42) has a unique solution $\varphi_{\alpha,\beta} \in \text{Lip } H \cap D(L_p)$ for any $p \geq 2$. Moreover*

$$\|\varphi_{\alpha,\beta}\|_{\text{Lip}} \leq \frac{1}{\lambda} \|f\|_1. \quad (11.44)$$

Finally $\varphi_{\alpha,\beta} \in D(N_2)$ and

$$N_2 \varphi_{\alpha,\beta} = \lambda \varphi_{\alpha,\beta} - f + \langle D_x U - D_x U_{\alpha,\beta}, D_x \varphi_{\alpha,\beta} \rangle. \quad (11.45)$$

Proof. We claim that the solution of (11.42) is given by

$$\varphi_{\alpha,\beta}(x) = \int_0^{+\infty} e^{-\lambda t} \mathbb{E}[f(X_{\alpha,\beta}(t,x))]dt, \quad x \in H, \ t \geq 0,$$

where $X_{\alpha,\beta}$ is the solution to (11.43). First notice that $\varphi_{\alpha,\beta}$ is of class C^1 and for $t > 0$, $x \in H$

$$\langle D_x \varphi_{\alpha,\beta}(x), h \rangle = \int_0^{+\infty} e^{-\lambda t} \mathbb{E}[\langle D_x f(X_{\alpha,\beta}(t,x)), D_x X_{\alpha,\beta}(t,x)h \rangle]dt.$$

Moreover, it is not difficult to see that $\varphi_{\alpha,\beta} \in D(L_p)$ for all $p \geq 2$ and

$$\lambda \varphi_{\alpha,\beta} - L_p \varphi_{\alpha,\beta} + \langle D_x U_{\alpha,\beta}, D_x \varphi_{\alpha,\beta} \rangle = f.$$

To go further, we need the following continuous inclusion,

$$L^4(H,\mu) \subset L^2(H,\nu).$$

We have in fact, by the Hölder inequality,

$$\left(\int_H \varphi^2 d\nu \right)^2 = \left(\int_H \varphi^2 \rho d\mu \right)^2 \leq \int_H \varphi^4 d\mu, \quad \varphi \in L^2(H,\nu).$$

We are now ready to show that $\varphi_{\alpha,\beta} \in D(N_2)$. Since $\varphi_\alpha \in D(L_4)$, and $\mathscr{E}_A(H)$ is a core for L_4, there exists a sequence $(\varphi_{\alpha,\beta,n}) \subset \mathscr{E}_A(H)$ such that

$$\lim_{n\to\infty} \varphi_{\alpha,\beta,n} = \varphi_{\alpha,\beta}, \ \lim_{n\to\infty} L_4 \varphi_{\alpha,\beta,n} = L_4 \varphi_\alpha \quad \text{in } L^4(H,\mu).$$

Since $D(L_4) \subset W^{1,4}(H,\mu)$ with continuous embedding (this fact is an obvious generalization of Proposition 10.22), we also have

$$\lim_{n\to\infty} D_x \varphi_{\alpha,\beta,n} = D_x \varphi_{\alpha,\beta} \quad \text{in } L^4(H,\mu;H).$$

Set

$$f_n = \lambda \varphi_{\alpha,\beta,n} - L_4 \varphi_{\alpha,\beta,n} + \langle D_x U_{\alpha,\beta}, D_x \varphi_{\alpha,\beta,n} \rangle.$$

We claim that

$$\lim_{n\to\infty} f_n = f \quad \text{in } L^2(H,\nu).$$

For this it is enough to show that

$$\lim_{n\to\infty} \int_H |\langle D_x U_{\alpha,\beta}, D_x \varphi_{\alpha,\beta} - D_x \varphi_{\alpha,\beta,n} \rangle|^2 d\nu = 0.$$

We have in fact, since there is $c_{\alpha,\beta} > 0$ such that $|D_x U_{\alpha,\beta}(x)| \leq c_{\alpha,\beta}$ $(1 + |x|)$, $x \in H$,

$$\int_H |\langle D_x U_{\alpha,\beta}, D_x \varphi_{\alpha,\beta} - D_x \varphi_{\alpha,\beta,n}\rangle|^2 d\nu \le \int_H |D_x U_{\alpha,\beta}|^2 |D_x \varphi_{\alpha,\beta}$$

$$-D_x \varphi_{\alpha,\beta,n}|^2 \rho d\mu \le c_{\alpha,\beta}^2 \int_H |D_x \varphi_{\alpha,\beta} - D_x \varphi_{\alpha,\beta,n}|^4 d\mu \int_H (1 + |x|^2 \rho^2) d\mu.$$

Since $f_n \to f$ in $L^2(H, \nu)$, it follows that $\varphi_{\alpha,\beta} \in D(N_2)$ and (11.45) holds.

Finally (11.44) follows from (11.36). $\square$

We are now ready to prove the main result of this chapter.

Theorem 11.24 N_2 *is self-adjoint in* $L^2(H, \nu)$.

Proof. We shall prove that for any $\lambda > 0$ the range of $\lambda - N_2$ is dense in $L^2(H, \nu)$.

Let $f \in C_b^1(H)$, $\alpha > 0$, $\beta > 0$ and let $\varphi_{\alpha,\beta}$ be the solution of (11.42). By Proposition 11.23 we have

$$\lambda \varphi_{\alpha,\beta} - N_2 \varphi_{\alpha,\beta} = f - \langle D_x U - D_x U_{\alpha,\beta}, D_x \varphi_{\alpha,\beta} \rangle. \qquad (11.46)$$

Taking into account (11.44) we find that

$$\int_H |\langle D_x U - D_x U_\alpha, D_x \varphi_{\alpha,\beta}\rangle|^2 d\nu \le \frac{1}{\lambda} \|f\|_1 \int_H |D_x U - D_x U_{\alpha,\beta}|^2 d\nu$$

$$\le \frac{1}{\lambda} \|f\|_1 \int_H |D_x U - D_x U_\alpha|^2 d\nu + \frac{1}{\lambda} \|f\|_1 \int_H |D_x U_\alpha - D_x U_{\alpha,\beta}|^2 d\nu$$

$$: = \frac{1}{\lambda} \|f\|_1 (I_\alpha + I_{\alpha,\beta}).$$

Since $D_x U_\alpha$ is Lipschitz continuous, there exists $c_\alpha > 0$ such that

$$|D_x U_{\alpha,\beta}(x)| + |D_x U_\alpha(x)| \le c_\alpha (1 + |x|), \quad x \in H.$$

Consequently,

$$|D_x U_\alpha(x) - D_x U_{\alpha,\beta}(x)|^2 \le 4 c_\alpha^2 (1 + |x|)^2, \quad x \in H,$$

and since

$$\int_H (1 + |x|)^2 \nu(dx) = \int_H (1 + |x|)^2 \rho(x) \mu(dx) < +\infty,$$

we can conclude, by the dominated convergence theorem, that

$$\lim_{\beta \to 0} I_{\alpha,\beta} = 0, \quad \alpha > 0.$$

Since, again by the dominated convergence theorem,

$$\lim_{\alpha \to 0} I_\alpha = 0,$$

we have proved that

$$\lim_{\alpha \to 0} \lim_{\beta \to 0} \langle D_x U - D_x U_{\alpha,\beta}, D_x \varphi_{\alpha,\beta} \rangle = 0 \quad \text{in } L^2(H, \nu).$$

This implies that the closure of $(\lambda - N_2)(L^2(H, \nu))$ includes $C_b^1(H)$ which is dense on $L^2(H, \nu)$. Therefore the range of $\lambda - N_2$ is dense in $L^2(H, \nu)$, and the conclusion follows from the Lumer–Phillips theorem (Theorem A.21). $\square$

By Proposition 11.16 it follows that

Proposition 11.25 *Assume that Hypotheses 11.1 and 11.2 hold. Then for all $\varphi, \psi \in D(N_2)$ we have*

$$\int_H N_2 \varphi \, \psi d\nu = -\frac{1}{2} \int_H \langle D_x \varphi, D_x \psi \rangle d\nu. \tag{11.47}$$

Proof. Let $\varphi, \psi \in D(N_2)$. Since $\mathscr{E}_A(H)$ is a core for N_2 there exist sequences $(\varphi_n), (\psi_n)$ in $\mathscr{E}_A(H)$ such that

$$\lim_{n \to \infty} \varphi_n = \varphi, \quad \lim_{n \to \infty} \psi_n = \psi \quad \text{in } L^2(H, \nu).$$

In view of (11.23) we have

$$\int_H N_0 \varphi_n \, \psi_n d\nu = -\frac{1}{2} \int_H \langle D_x \varphi_n, D_x \psi_n \rangle d\nu.$$

Now, the conclusion follows, letting $n \to \infty$. $\square$

We shall denote by P_t the semigroup generated by N_2. We can prove now an interesting identity.

Proposition 11.26 *Assume that Hypotheses 11.1 and 11.2 hold. Then for all $\varphi \in D(N_2)$ we have*

$$\int_{\mathbb{R}^d} (P_t \varphi)^2 \, \nu(dx) + \int_0^t ds \int_{\mathbb{R}^d} |D_x P_s \varphi|^2 \, \nu(dx) = \int_{\mathbb{R}^d} \varphi^2 \, \nu(dx). \tag{11.48}$$

Proof. Let $\varphi \in D(N_2)$ and write

$$\frac{1}{2} \frac{d}{dt} (P_t \varphi)^2 = N_2 P_t \varphi \, P_t \varphi.$$

Integrating this identity over H with respect to ν and taking into account (11.47), yields

$$\frac{1}{2}\frac{d}{dt}\int_H |P_t\varphi|^2 d\nu = \int_H N_2 P_t\varphi \, P_t\varphi \, d\nu = -\frac{1}{2}\int_H |D_x P_t\varphi|^2 d\nu.$$

Now, (11.47) follows, integrating with respect to t. $\square$

11.7 Asymptotic behaviour of P_t

Proposition 11.27 *For any $\varphi \in L^2(H,\nu)$ we have*

$$\lim_{n\to\infty} P_t\varphi = \int_H \varphi d\nu \quad \text{in } L^2(H,\nu). \tag{11.49}$$

Proof. Let $\varphi \in L^2(H,\nu)$. By (11.48) we deduce that the function

$$[0,\infty) \to \mathbb{R}, \; t \mapsto \int_H (P_t\varphi)^2 d\nu$$

is decreasing. So, there exists $\gamma(\varphi) \geq 0$ such that

$$\lim_{t\to+\infty} \|P_t\varphi\|^2_{L^2(H,\nu)} = \gamma(\varphi).$$

It follows that

$$\lim_{t\to\infty} \langle P_t\varphi, P_t\psi \rangle_{L^2(H)} = \frac{1}{2}\left[\gamma(\varphi+\psi) - \gamma(\varphi) - \gamma(\psi)\right] \quad \text{for all } \varphi,\psi \in L^2(H,\nu).$$

This implies, in view of the symmetry of P_t, that there exists the limit

$$\lim_{t\to\infty} \langle P_t\varphi, \psi \rangle_{L^2(H,\nu)} \quad \text{for all } \varphi,\psi \in L^2(H,\nu).$$

By the uniform boundedness theorem, there exists a bounded operator G in $L^2(H,\nu)$ and it is easy to see that

$$\lim_{t\to\infty} P_t\varphi = G\varphi \quad \text{for all } \varphi \in L^2(H,\nu). \tag{11.50}$$

Now, let $\varphi \in D(N_2)$. Then from (11.48) it follows that

$$\int_0^{+\infty} \|D_x P_t\varphi\|^2_{L^2(H,\nu)} ds \leq \|\varphi\|^2_{L^2(H,\nu)}, \tag{11.51}$$

and

$$\int_0^{+\infty} \|D_x P_t N_2\varphi\|^2_{L^2(H,\nu)} ds \leq \|\varphi\|^2_{L^2(H,\nu)}. \tag{11.52}$$

Set

$$\alpha(t) = \|D_x P_t \varphi\|_{L^2(H,\nu)}^2, \quad \beta(t) = \|D_x P_t N_2 \varphi\|_{L^2(H,\nu)}^2, \quad t \geq 0.$$

Then we have

$$\alpha'(t) = 2\langle \alpha(t), \beta(t) \rangle, \quad t \geq 0.$$

Therefore

$$\alpha(t) \leq \|\varphi\|_{L^2(H,\nu)}^2, \quad |\alpha'(t)| \leq 2\|\varphi\|_{L^2(H,\nu)}^2 \|N_2\varphi\|_{L^2(H,\nu)}^2,$$

so that $\alpha \in W^{1,1}(0, +\infty)$. Consequently,

$$\lim_{t\to\infty} \alpha(t) = 0. \tag{11.53}$$

Finally, let $t_n \uparrow +\infty$ and $\psi \in L^2(H,\nu)$ such that $P_{t_n}\varphi \to \psi$ weakly. So,

$$P_{t_n}\varphi \rightharpoonup \psi, \quad D_x P_{t_n}\varphi \to 0.$$

This implies that ψ is constant and

$$\psi = \int_H \varphi(y)\nu(dy).$$

The conclusion follows. $\square$

From Proposition 11.27 it follows that the measure ν is *ergodic* and *strongly mixing*.

11.7.1 Poincaré and log-Sobolev inequalities

We assume here that Hypotheses 11.1 and 11.2 hold and that U is convex. For any $\alpha > 0$ and $\beta > 0$ we denote by $\nu_{\alpha,\beta}$ the probability measure

$$\nu_{\alpha,\beta}(dx) = Z_{\alpha,\beta}^{-1} e^{-2U_{\alpha,\beta}(x)} \mu(dx),$$

where

$$Z_{\alpha,\beta} = \int_H e^{-2U_{\alpha,\beta}(y)} \mu(dy),$$

and $U_{\alpha,\beta}$ is defined by (11.38).

By the dominated convergence theorem we have

$$\lim_{\alpha\to0}\lim_{\beta\to0} \nu_{\alpha,\beta} = \nu \quad \text{weakly.} \tag{11.54}$$

Moreover, we denote by $X_{\alpha,\beta}(t,x)$ the solution of (11.43) and set $\eta_{\alpha,\beta}^h(t,x) = D_x X_{\alpha,\beta}(t,x)$. By (11.34) it follows that for any $h \in H$, we have

$$|\eta_{\alpha,\beta}^h(t,x)| \leq e^{-\omega t}|h|, \quad x \in H, \ t \geq 0. \tag{11.55}$$

Lemma 11.28 *For all $\varphi \in W^{1,2}(H, \nu_{\alpha,\beta})$ we have*

$$\int_H |\varphi - \overline{\varphi}_{\alpha,\beta}|^2 d\nu_{\alpha,\beta} \leq \frac{1}{2\omega} \int_H |D_x\varphi|^2 d\nu_{\alpha,\beta}, \tag{11.56}$$

where

$$\overline{\varphi}_{\alpha,\beta} = \int_H \varphi d\nu_{\alpha,\beta}.$$

and

$$\int_H \varphi^2 \log(\varphi^2) d\nu_{\alpha,\beta} \leq \frac{1}{\omega} \int_H |D_x\varphi|^2 d\nu_{\alpha,\beta} + \|\varphi\|_{L^2(H,\nu_{\alpha,\beta})}^2 \log(\|\varphi\|_{L^2(H,\nu_{\alpha,\beta})}^2). \tag{11.57}$$

Proof. To prove (11.56) we proceed as for the proof of Theorem 10.25, using the basic ingredients (11.49) and (11.55). Similarly, to prove (11.57) we proceed as in the proof of Theorem 10.30. $\square$

Finally, letting $\alpha, \beta \to 0$ and invoking (11.54) we find the following results.

Proposition 11.29 *Assume that Hypotheses 11.1 and 11.2 hold and U is convex. Then, for any $\varphi \in W^{1,2}(H, \nu)$ we have*

$$\int_H |\varphi - \overline{\varphi}|^2 \, d\nu \leq \frac{1}{2\omega} \int_H |D_x\varphi|^2 d\mu, \tag{11.58}$$

where $\overline{\varphi} = \int_H \varphi d\nu$ and

$$\int_H \varphi^2 \log(\varphi^2) d\nu \leq \frac{1}{\omega} \int_H |D_x\varphi|^2 d\nu + \|\varphi\|_{L^2(H,\nu)}^2 \log(\|\varphi\|_{L^2(H,\nu)}^2). \tag{11.59}$$

By Proposition 11.29 (arguing as in the proof of Propositions 10.27 and 10.28) we obtain a *spectral gap* of N_2 and exponential convergence to equilibrium for P_t.

Corollary 11.30 *Under the assumptions of Proposition 11.29 we have*

$$\sigma(N_2) \subset \{\lambda \in \mathbb{C} : \Re \lambda \leq \omega\},$$

where $\sigma(N_2)$ is the spectrum of N_2. Moreover,

$$\int_H |P_t\varphi - \overline{\varphi}|^2 d\nu \leq e^{-2\omega t} \int_H \varphi^2 d\nu, \quad t \geq 0.$$

Finally, with the same proof as Theorem 10.31, we obtain the hyper-contractivity of P_t.

Theorem 11.31 *Assume that Hypotheses 11.1 anrd 11.2 hold and U is convex. Then for all $t > 0$ we have*

$$\|R_t\varphi\|_{L^{q(t)}(H,\mu)} \le \|\varphi\|_{L^p(H,\mu)}, \quad p \ge 2, \ \varphi \in L^p(H,\mu), \tag{11.60}$$

where

$$q(t) = 1 + (p-1)e^{2\omega t}, \quad t > 0. \tag{11.61}$$

11.7.2 Compactness of the embedding of $W^{1,2}(H,\nu)$ in $L^2(H,\nu)$

In this section we follow [8].

Theorem 11.32 *Assume that Hypotheses 11.1 and 11.2 are fulfilled and, in addition, that there exists $\varepsilon \in [0,1]$ such that*

$$\int_H |D_x \log \rho|^{2+\varepsilon} d\nu < +\infty. \tag{11.62}$$

Then the embedding $W^{1,2}(H,\nu) \subset L^2(H,\nu)$ is compact.

Proof. Let $(\varphi_n) \subset W^{1,2}(H,\nu)$ be such that

$$\int_H |\varphi_n|^2 d\nu + \int_H |D_x\varphi_n|^2 d\nu \le 1, \quad n \in \mathbb{N}.$$

We have to show that there exists a subsequence (φ_{n_k}) which is convergent in $L^2(H,\nu)$. Notice that, in view of the log-Sobolev inequality (11.59), the sequence (φ_n^2) is uniformly integrable. Thus it is enough to find a subsequence (φ_{n_k}) which is pointwise convergent, since then, by the Vitali theorem, one can conclude that (φ_{n_k}) is convergent in $L^2(H,\nu)$.

To construct a pointwise convergent subsequence, we proceed as follows. First for any $R > 1$ we set

$$G_R = \left\{ x \in H : \rho(x) \ge \frac{1}{R} \right\},$$

and notice that $\lim_{R\to\infty} \nu(G_R) = 1$. Then we consider a function $\theta : [0,+\infty) \to [0,+\infty)$ of class C^∞ such that

$$\theta(r) \begin{cases} = 1 & \text{if } r \in [0,1] \\ = 0 & \text{if } r \ge 2 \\ \in [0,1] & \text{if } r \in [1,2], \end{cases}$$

and set

$$\varphi_{n,R}(x) = \theta\left(\frac{-2\log \rho(x)}{\log R}\right) \varphi_n(x), \quad x \in H,$$

so that

$$\varphi_{n,R}(x) \begin{cases} \leq |\varphi_n(x)|, & x \in H \\ = 0 & \text{if } \rho(x) < \frac{1}{R} \\ = \varphi_n(x) & \text{if } \rho(x) \geq \frac{1}{\sqrt{R}}. \end{cases}$$

Finally, we prove that there exists $C_R > 0$ and $\alpha \in (0,1)$ such that

$$\int_H |\varphi_{n,R}|^2 d\mu + \int_H |D_x \varphi_{n,R}|^{1+\alpha} d\mu \leq C_R. \tag{11.63}$$

Once the claim is proved, since the embedding $W^{1,1+\alpha}(H,\mu) \subset L^{1+\alpha}(H,\mu)$ is compact, see [5], we can construct a subsequence $(\varphi_{n_k,R})$ which is convergent in $L^{1+\alpha}(H,\nu)$ and then another subsequence which is pointwise convergent. Now, by a standard diagonal procedure we can find a subsequence (φ_{n_k}) pointwise convergent as required.

It remains to show (11.63). We have in fact

$$\int_H |\varphi_{n,R}|^2 d\mu = \int_{\{\rho \geq \frac{1}{R}\}} |\varphi_{n,R}|^2 \frac{d\nu}{\rho} \leq R \int_H |\varphi_n|^2 d\mu \leq R. \tag{11.64}$$

Moreover

$$D_x \varphi_{n,R}(x) = -2\theta' \left(\frac{-2\log \rho(x)}{\log R} \right) \left(\frac{D_x \log \rho(x)}{\log R} \right) \varphi_n(x)$$

$$+ \theta \left(\frac{-2\log \rho(x)}{\log R} \right) D_x \varphi_n(x) := F_{n,R}(x) + H_{n,R}(x)$$

and

$$\int_H |H_{n,R}|^2 d\mu \leq \int_{\{\rho \geq \frac{1}{R}\}} |D_x \varphi_{n,R}|^2 \frac{d\nu}{\rho} \leq R \int_H |D_x \varphi_n|^2 d\mu \leq R. \tag{11.65}$$

Also, for any $\alpha \in (0,1)$ we have, using the Hölder inequality

$$\int_H |F_{n,R}|^{1+\alpha} d\mu \leq \left(\frac{2}{\log R} \right)^{1+\alpha} R \int_H |D \log \rho|^{1+\alpha} |\varphi_n|^{1+\alpha} d\mu$$

$$\leq \left(\frac{2}{\log R} \right)^{1+\alpha} R \left(\int_H |D \log \rho|^{\frac{2(1+\alpha)}{1-\alpha}} d\nu \right)^{\frac{1-\alpha}{2}} \left(\int_H |\varphi_n|^2 d\nu \right)^{\frac{1+\alpha}{2}}.$$
$$\tag{11.66}$$

Choosing $\alpha = \frac{\varepsilon}{4+\varepsilon}$ we have $\frac{2(1+\alpha)}{1-\alpha} = 2+\varepsilon$. Then by (11.65) and (11.66), (11.63) follows. $\square$

Exercise 11.33 Consider the motivating example in section 11.2. Show that in this case the assumption (11.62) is fulfilled.

A

Linear semigroups theory

Throughout this appendix X represents a complex [1] Banach space (norm $|\cdot|$), and $L(X)$ the Banach algebra of all linear bounded operators from X into X endowed with the sup norm

$$\|T\| = \sup\{|Tx| : x \in X, \, |x| \leq 1\}.$$

A.1 Some preliminaries on spectral theory

A.1.1 Closed and closable operators

Let $T: D(T) \subset X \to X$ be a linear operator. The *graph* of T is defined by

$$\mathscr{G}_T = \{(x, y) \in X \times X : x \in D(T), \, y = Tx\},$$

where $X \times X$ denotes the product Banach space endowed with the norm

$$|(x, y)| := |x| + |y|, \quad (x, y) \in X \times Y.$$

The operator T is said to be *closed* if its graph $\mathscr{G}_T$ is a closed subset of $X \times X$.

It is easy to check that T is closed if and only if for any sequence $(x_n) \subset D(T)$ such that

$$\lim_{n \to \infty} x_n \to x, \quad \lim_{n \to \infty} Tx_n \to y,$$

where $y \in X$, one has $x \in D(T)$ and $y = Tx$.

If T is closed, we endow $D(T)$ with the *graph norm* setting

$$|x|_{D(T)} = |x| + |Tx|, \quad x \in D(T).$$

---

[1] All results of this section have a natural extension to real Banach spaces.

It is easy to see that, since T is closed, $D(T)$ is a Banach space.

The operator T is said to be *closable* if there exists a linear operator $\overline{T}$ from X into X (necessarily unique) whose graph coincides with the closure of $\mathcal{G}_T$. If T is closable we call $\overline{T}$ the *closure* of T.

It is easy to check that T is closable if and only if for any sequence $(x_n) \subset D(T)$ such that

$$\lim_{n \to \infty} x_n \to 0, \quad \lim_{n \to \infty} Tx_n \to y,$$

where $y \in X$, one has $y = 0$.

Let $A \colon D(A) \subset X \to X$ be a linear closed operator. We say that $\lambda \in \mathbb{C}$ belongs to the *resolvent set* $\rho(A)$ of A if $\lambda - A$ is bijective and $(\lambda - A)^{-1} \in L(X)$; in this case the operator $R(\lambda, A) \colon = (\lambda - A)^{-1}$ is called the *resolvent* of A at λ. The complement $\sigma(A)$ of $\rho(A)$ is called the *spectrum* of A.

Example A.1 Let $X = C([0,1])$ be the Banach space of all continuous functions on $[0,1]$ endowed with the sup norm and let $C^1([0,1])$ be the subspace of $C([0,1])$ of all functions u which are continuously differentiable. Let us consider the following linear operator A on X,

$$\begin{cases} Au = u', & u \in D(A), \\ D(A) = C^1([0,1]). \end{cases}$$

Then,

$$\rho(A) = \varnothing, \ \sigma(A) = \mathbb{C}.$$

In fact, given $\lambda \in \mathbb{C}$, the mapping $\lambda - A$ is not injective since for all $c \in \mathbb{C}$ the function $u(\xi) = ce^{\lambda \xi}$ belongs to $D(A)$ and $(\lambda - A)u = 0$.

Consider now the linear operator,

$$\begin{cases} Bu = u', & u \in D(A), \\ D(B) = \{u \in C^1([0,1]); \ u(0) = 0\}. \end{cases}$$

Then we have

$$\rho(B) = \mathbb{C}, \ \sigma(A) = \varnothing$$

and

$$(R(\lambda, B)f)(\xi) = -\int_0^\xi e^{\lambda(\xi - \eta)} f(\eta) d\eta, \quad \lambda \in \mathbb{C}, \ f \in X, \ \xi \in [0,1].$$

In fact $\lambda \in \rho(B)$ if and only if the problem

$$\begin{cases} \lambda u(\xi) - u'(\xi) = f(\xi) \\ u(0) = 0 \end{cases}$$

has a unique solution $f \in X$.

Let us prove the *resolvent identity*.

Proposition A.2 *Let A be a closed operator. If $\lambda, \mu \in \rho(A)$ we have*

$$R(\lambda, A) - R(\mu, A) = (\mu - \lambda)R(\lambda, A)R(\mu, A). \qquad (A.1)$$

Proof. For all $x \in X$ we have

$$(\mu - \lambda)R(\lambda, A)x = (\mu - A + A - \lambda)R(\lambda, A)x = (\mu - A)R(\lambda, A)x - x.$$

Applying $R(\mu, A)$ to both sides of the above identity, we find

$$(\mu - \lambda)R(\mu, A)R(\lambda, A)x = R(\lambda, A)x - R(\mu, A)x$$

and the conclusion follows. $\square$

Proposition A.3 *Let A be a closed operator. Let $\lambda_0 \in \rho(A)$ and let λ be such that $|\lambda - \lambda_0| < \frac{1}{\|R(\lambda_0, A)\|}$. Then $\lambda \in \rho(A)$ and*

$$R(\lambda, A) = R(\lambda_0, A)(1 + (\lambda - \lambda_0)R(\lambda_0, A))^{-1}. \qquad (A.2)$$

Thus $\rho(A)$ is open and $\sigma(A)$ is closed. Moreover

$$R(\lambda, A) = \sum_{k=1}^{\infty}(-1)^k(\lambda - \lambda_0)^k R^{k+1}(\lambda_0, A), \qquad (A.3)$$

and so $R(\lambda, A)$ is analytic on $\rho(A)$.

Proof. Let $\lambda_0 \in \rho(A)$ and $|\lambda - \lambda_0| < \frac{1}{\|R(\lambda_0, A)\|}$. Equation $\lambda x - Ax = y$ is equivalent to

$$(\lambda - \lambda_0)x + (\lambda_0 - A)x = y,$$

and, setting $z = (\lambda_0 - A)x$, to

$$z + (\lambda - \lambda_0)R(\lambda_0, A)z = y.$$

Since $\|(\lambda - \lambda_0)R(\lambda_0, A)\| < 1$ it follows that

$$z = (1 + (\lambda - \lambda_0)R(\lambda_0, A))^{-1}y,$$

which yields the conclusion. $\square$

A.2 Strongly continuous semigroups

A *strongly continuous semigroup* of linear operators on X is a mapping

$$T: [0, \infty) \to L(X), \; t \mapsto T(t)$$

such that

(i) $T(t + s) = T(t)T(s)$ for all $t, s \geq 0$, $T(0) = I$.
(ii) $T(\cdot)x$ is continuous for all $x \in X$.

Remark A.4 Assume that T is a strongly continuous semigroup. Then $\|T(\cdot)\|$ is locally bounded by the uniform boundedness theorem.

The *infinitesimal generator* A of $T(\cdot)$ is defined by

$$\begin{cases} D(A) = \left\{ x \in X : \exists \lim_{h \to 0+} \Delta_h x \right\} \\ Ax = \lim_{h \to 0+} \Delta_h x, \end{cases} \tag{A.4}$$

where

$$\Delta_h = \frac{T(h) - I}{h}, \quad h > 0.$$

Proposition A.5 *Let T be a strongly continuous semigroup and let A be its infinitesimal generator. Then $D(A)$ is dense in X.*

Proof. For all $x \in H$ and $a > 0$ we set

$$x_a = \frac{1}{a} \int_0^a T(s)x\,ds.$$

Since $\lim_{a \to 0} x_a = x$, it is enough to show that $x_a \in D(A)$. We have in fact for any $h \in (0, a)$,

$$\Delta_h x_a = \frac{1}{ah} \left[\int_a^{a+h} T(s)x\,ds - \int_0^h T(s)x\,ds \right],$$

and, consequently $x_a \in D(A)$ since

$$\lim_{h \to 0} \Delta_h x_a = \Delta_a x.$$

$\square$

Exercise A.6 *Prove that $D(A^2)$ is dense in X.*

We now study the derivability of the semigroup $T(t)$. Let us first notice that, since $\Delta_h T(t)x = T(t)\Delta_h x$, if $x \in D(A)$ then $T(t)x \in D(A)$ for all $t \geq 0$ and $AT(t)x = T(t)Ax$.

Proposition A.7 *Assume that $x \in D(A)$, then $T(\cdot)x$ is differentiable for all $t \geq 0$ and*

$$\frac{d}{dt} T(t)x = AT(t)x = T(t)Ax. \tag{A.5}$$

Proof. Let $t_0 \geq 0$ be fixed and let $h > 0$. Then we have

$$\frac{T(t_0 + h)x - T(t_0)x}{h} = \Delta_h T(t_0)x \overset{h \to 0}{\to} AT(t_0)x.$$

This shows that $T(\cdot)x$ is right differentiable at t_0. Let us show left differentiability, assuming $t_0 > 0$. For $h \in (0, t_0)$ we have

$$\frac{T(t_0 - h)x - T(t_0)x}{h} = T(t_0 - h)\Delta_h x \overset{h \to 0}{\to} T(t_0)Ax,$$

since $\|T(t)\|$ is locally bounded by Remark A.4. $\square$

Proposition A.8 *Let T be a strongly continuous semigroup and let A be its infinitesimal generator. Then A is a closed operator.*

Proof. Let $(x_n) \subset D(A)$, and let $x, y \in X$ be such that

$$x_n \to x, \quad Ax_n = y_n \to y$$

Then we have

$$\Delta_h x_n = \frac{1}{h} \int_0^h T(t)y_n dt.$$

As $h \to 0$ we get $x \in D(A)$ and $y = Ax$, so that A is closed. $\square$

We end this section by studying the asymptotic behaviour of $T(\cdot)$. We define the *type* of $T(\cdot)$ by

$$\omega_0 = \inf_{t>0} \frac{\log \|T(t)\|}{t}.$$

Clearly $\omega_0 \in [-\infty, +\infty)$.

Proposition A.9 *We have*

$$\omega_0 = \lim_{t \to +\infty} \frac{1}{t} \log \|T(t)\|. \qquad (A.6)$$

Proof. It is enough to show that

$$\limsup_{t \to \infty} \frac{1}{t} \log \|T(t)\| \leq \omega_0.$$

Let $\varepsilon > 0$ and $t_\varepsilon > 0$ be such that

$$\frac{1}{t_\varepsilon} \log \|T(t_\varepsilon)\| < \omega_0 + \varepsilon.$$

Set
$$t = n(t)t_\varepsilon + r(t), \quad n(t) \in \mathbb{N}, \quad r(t) \in [0, t_\varepsilon).$$

Since $\|T(\cdot)\|$ is locally bounded (see Remark A.4), there exists $M_\varepsilon > 0$ such that
$$\|T(t)\| \le M_\varepsilon \quad \text{for all } t \in [0, t_\varepsilon].$$

Write
$$\frac{1}{t} \log \|T(t)\| = \frac{1}{t} \log \|T(t_\varepsilon)^{n(t)} T(r(t))\|$$

$$\le \frac{1}{n(t)t_\varepsilon + r(t)} \left(n(t) \log \|T(t_\varepsilon)\| + \log \|T(r(t))\| \right)$$

$$\le \frac{1}{t_\varepsilon + \frac{r(t)}{n(t)}} \left(\log \|T(t_\varepsilon)\| + \frac{M_{t_\varepsilon}}{n(t)} \right).$$

As $t \to +\infty$, we obtain
$$\limsup_{t\to\infty} \frac{1}{t} \log \|T(t)\| \le \frac{1}{t_\varepsilon} \log \|T(t_\varepsilon)\| \le \omega_0 + \varepsilon.$$

$\square$

Corollary A.10 *Let T be a strongly continuous semigroup of type ω_0. Then for all $\varepsilon > 0$ there exists $N_\varepsilon \ge 1$ such that*
$$\|T(t)\| \le N_\varepsilon e^{(\omega_0+\varepsilon)t}, \quad t \ge 0. \tag{A.7}$$

Proof. Let $t_\varepsilon, n(t), r(t)$ as in the previous proof. Then we have
$$\|T(t)\| \le \|T(t_\varepsilon)\|^{n(t)} \|T(r(t))\| \le e^{t_\varepsilon n(t)(\omega_0+\varepsilon)} M_{t_\varepsilon} \le M_{t_\varepsilon} e^{(\omega_0+\varepsilon)t}$$

and the conclusion follows. $\square$

We shall denote by $\mathscr{G}(M, \omega)$ the set of all strongly continuous semigroups T such that
$$\|T(t)\| \le M e^{\omega t}, \quad t \ge 0.$$

Example A.11 Let $X = L^p(\mathbb{R}), p \ge 1, (T(t)f)(\xi) = f(\xi - t), f \in L^p(\mathbb{R})$. Then we have $\|T(t)\| = 1$ and so $\omega_0 = 0$.

Example A.12 Let $X = L^p(0, T), T > 0, p \ge 1$, and let
$$(T(t)f)(\xi) = \begin{cases} f(\xi - t) & \text{if } \xi \in [t, T] \\ \\ 0 & \text{if } \xi \in [0, t). \end{cases}$$

Then we have $T(t) = 0$ if $t \ge T$ and so $\omega_0 = -\infty$.

Exercise A.13 Let $A \in L(X)$ be a compact operator and let $(\lambda_i)_{i \in \mathbb{N}}$ be the set of its eigenvalues. Set $T(t) = e^{tA}$. Prove that

$$\omega_0 = \sup_{i \in \mathbb{N}} \Re \lambda_i.$$

A.3 The Hille–Yosida theorem

Proposition A.14 *Let T be a strongly continuous semigroup belonging to $G(M, \omega)$ and let A its infinitesimal generator. Then we have*

$$\rho(A) \supset \{\lambda \in \mathbb{C}, \ \Re \lambda > \omega\}, \tag{A.8}$$

$$R(\lambda, A)y = \int_0^\infty e^{-\lambda t} T(t) y \, dt, \quad y \in X, \ \Re \lambda > \omega. \tag{A.9}$$

Proof. Set

$$\Sigma = \{\lambda \in \mathbb{C}; \ \Re \lambda > \omega\},$$

$$F(\lambda)y = \int_0^\infty e^{-\lambda t} T(t) y \, dt, \ y \in X, \ \Re \lambda > \omega.$$

We notice that the function $t \mapsto e^{-\lambda t} T(t) y$ is integrable since $T \in \mathscr{G}(M, \omega)$. We have to show that, given $\lambda \in \Sigma$ and $y \in X$, the equation $\lambda x - Ax = y$ has a unique solution given by $x = F(\lambda)y$.

Existence. Let $\lambda \in \Sigma, y \in X, \ x = F(\lambda)y$. Then we have

$$\Delta_h x = \frac{1}{h}(e^{\lambda h} - 1)x - \frac{1}{h} e^{\lambda h} \int_0^h e^{-\lambda t} T(t) y \, dt$$

and so, as $h \to 0$,

$$\lim_{h \to 0+} \Delta_h x = \lambda x - y = Ax$$

so that x is a solution of the equation $\lambda x - Ax = y$.

Uniqueness. Let $x \in D(A)$ be a solution of the equation $\lambda x - Ax = y$. Then we have

$$\int_0^\infty e^{-\lambda t} T(t)(\lambda x - Ax) dt = \lambda \int_0^\infty e^{-\lambda t} T(t) x \, dt$$

$$- \int_0^\infty e^{-\lambda t} \frac{d}{dt} T(t) x \, dt = x,$$

so that $x = F(\lambda)y$. $\square$

We are now going to prove the *Hille–Yosida* theorem.

Theorem A.15 *Let* $A: D(A) \subset X \to X$ *be a closed operator. Then* A *is the infinitesimal generator of a strongly continuous semigroup belonging to* $\mathscr{G}(M, \omega)$ *if and only if*

(i) $\rho(A) \supset \{\lambda \in \mathbb{R};\ \lambda > \omega\}.$

(ii) $\|R^n(\lambda, A)\| \leq \dfrac{M}{(\lambda - \omega)^n}$ *for all* $n \in \mathbb{N},\ \lambda > \omega.$

(iii) $D(A)$ *is dense in* $X.$ (A.10)

Given a linear operator A fulfilling (A.10) it is convenient to introduce a sequence of linear operators (called the *Yosida approximations* of A). They are defined as

$$A_n = nAR(n, A) = n^2 R(n, A) - n. \tag{A.11}$$

Lemma A.16 *We have*

$$\lim_{n \to \infty} nR(n, A)x = x \quad \text{for all } x \in X, \tag{A.12}$$

and

$$\lim_{n \to \infty} A_n x = Ax \quad \text{for all } x \in D(A). \tag{A.13}$$

Proof. Since $D(A)$ is dense in X and $\|nR(n, A)\| \leq \frac{Mn}{n-\omega}$, to prove (A.12) it is enough to show that

$$\lim_{n \to \infty} nR(n, A)x = x \quad \text{for all } x \in D(A).$$

In fact for any $x \in D(A)$ we have

$$|nR(n, A)x - x| = |R(n, A)Ax| \leq \frac{M}{n - \omega}|Ax|,$$

and the conclusion follows.

Finally if $x \in D(A)$ we have

$$A_n x = nR(n, A)Ax \to Ax,$$

and (A.13) follows. $\square$

    Proof of Theorem A.15. *Necessity.* (i) follows from Proposition A.14 and (iii) from Proposition A.5. Let us show (ii). Let $k \in \mathbb{N}$ and $\lambda > \omega$. Write

$$\frac{d^k}{d\lambda^k}R(\lambda, A)y = \int_0^\infty (-t)^k e^{-\lambda t}T(t)y\,dt, \quad y \in X,$$

then

$$\left\|\frac{d^k}{d\lambda^k}R(\lambda, A)\right\| \leq M\int_0^\infty t^k e^{-\lambda t+\omega t}\,dt$$

which yields the conclusion.

Sufficiency. **Step 1**. We have

$$\|e^{tA_n}\| \leq M e^{\frac{\omega n t}{n-\omega}} \quad \text{for all } n \in \mathbb{N}. \tag{A.14}$$

In fact, by the identity

$$e^{tA_n} = e^{-nt}e^{tn^2 R(n,A)} = e^{-nt}\sum_{k=0}^\infty \frac{n^{2k}t^k R^k(n, A)}{k!},$$

it follows that

$$\|e^{tA_n}\| \leq M e^{-nt}\sum_{k=0}^\infty \frac{n^{2k}t^k}{(n-\omega)^k k!}.$$

Step 2. There exists $C > 0$ such that, for all $m, n > 2\omega$, and $x \in D(A^2)$,

$$\|e^{tA_n}x - e^{tA_m}x\| \leq Ct\,\frac{|m-n|}{(m-\omega)(n-\omega)}\,\|A^2 x\|. \tag{A.15}$$

Setting $u_n(t) = e^{tA_n}x$, we find that

$$\frac{d}{dt}(u_n(t) - u_m(t)) = A_n(u_n(t) - u_m(t)) - (A_m - A_n)u_m(t)$$

$$= A_n(u_n(t) - u_m(t)) - (n-m)A^2 R(m, A)R(n, A)u_m(t).$$

It follows that

$$u_n(t) - u_m(t) = (n-m)A^2 R(m, A)R(n, A)\int_0^t e^{(t-s)A_n}u_m(s)\,ds$$

$$= (n-m)R(m, A)R(n, A)\int_0^t e^{(t-s)A_n}e^{sA_m}A^2 x.$$

Step 3. For all $x \in X$ there exists the limit

$$\lim_{n \to \infty} e^{tA_n} x =: T(t)x \qquad \text{(A.16)}$$

and the semigroup $T \colon [0, \infty) \to L(X), t \mapsto T(t)$ is strongly continuous.

From the second step it follows that the sequence $(u_n(t))$ is Cauchy, uniformly in t on compact subsets of $[0, +\infty)$, for all $x \in D(A^2)$. Since $D(A^2)$ is dense in X (see Exercise A.6) this holds for all $x \in X$. Finally, strong continuity of $T(\cdot)$ is easy to check.

Step 4. $T(\cdot)x$ is differentiable for all $x \in D(A)$ and

$$\frac{d}{dt} T(t)x = T(t)Ax = AT(t)x.$$

In fact let $x \in D(A)$, and $v_n(t) = \frac{d}{dt} u_n(t)$. Then

$$v_n(t) = e^{tA_n} A_n x.$$

Since $x \in D(A)$ there exists the limit

$$\lim_{n \to \infty} v_n(t) = e^{tA} Ax.$$

This implies that u is differentiable and $u'(t) = v(t)$ so that $u \in C^1([0, +\infty); X)$. Moreover

$$A(nR(n, A)u_n(t)) = u_n'(t) \to v(t).$$

Since A is closed and $nR(n, A)u_n(t) \to u(t)$ as $n \to \infty$ it follows that $u(t) \in D(A)$ and $u'(t) = Au(t)$.

Step 5. A is the infinitesimal generator of $T(\cdot)$.

Let B be the infinitesimal generator of $T(\cdot)$. By Step 4, $B \supset A$.[2] It is enough to show that if $x \in D(B)$ then $x \in D(A)$. Let $x \in D(B)$, $\lambda_0 > \omega$, setting $z = \lambda_0 x - Bx$ we have

$$z = (\lambda_0 - A)R(\lambda_0, A)z$$

$$= \lambda_0 R(\lambda_0, A)z - BR(\lambda_0, A)z = (\lambda_0 - B)R(\lambda_0, A)z.$$

Thus $x = R(\lambda_0, B)z = R(\lambda_0, A)z \in D(A)$. $\square$

---

[2] That is $D(B) \supset D(A)$ and $Ax = Bx$ for all $x \in D(A)$.

Remark A.17 To use the Hille–Yosida theorem requires to check infinite conditions. However if $M = 1$ it is enough to prove (ii) only for $n = 1$. If $\omega \leq 0$ we say that $T(\cdot)$ is a *contraction semigroup*.

Example A.18 Let $X = C_0([0, \pi])$ the Banach space of all continuous functions in $[0, \pi]$ which vanish at 0 and π endowed with the sup norm. Let A be the linear operator in X defined as

$$
\begin{cases}
D(A) = \{y \in C^2([0, \pi]); y(0) = y''(0) = y(\pi) = y''(\pi) = 0\} \\[2mm]
Ay = y'', \quad y \in D(A).
\end{cases}
$$

It is easy to check that $\sigma(A) = \{-n^2; \, n \in \mathbb{N}\}$. Moreover any element of $\sigma(A)$ is a simple eigenvalue whose corresponding eigenvector is given by

$$
\varphi_n(\xi) = \sin n\xi \quad \text{for all } n \in \mathbb{N},
$$

that is

$$
A\varphi_n = -n^2 \varphi_n \quad \text{for all } n \in \mathbb{N}.
$$

Moreover if $\lambda \in \rho(A)$ and $f \in C_0([0, \pi])$, $u = R(\lambda, A)f$ is the solution of the problem

$$
\begin{cases}
\lambda u(\xi) - u''(\xi) = f(\xi) \\[2mm]
u(0) = u(\pi) = 0.
\end{cases}
$$

By a direct verification we find that

$$
u(\xi) = \frac{\sinh(\sqrt{\lambda}\xi)}{\sqrt{\lambda}\sinh(\sqrt{\lambda}\pi)} \int_\xi^\pi \sinh[\sqrt{\lambda}(\pi - \eta)]f(\eta)d\eta
$$

$$
+ \frac{\sinh[\sqrt{\lambda}(\pi - \xi)]}{\sqrt{\lambda}\sinh(\sqrt{\lambda}\pi)} \int_0^\xi \sinh[\sqrt{\lambda}\eta]f(\eta)d\eta. \tag{A.17}
$$

From (A.17) it follows by a straightforward computation that

$$
\|R(\lambda, A)\| \leq \frac{1}{\lambda}, \quad \text{for all } \lambda > 0. \tag{A.18}
$$

Therefore all assumptions of the Hille–Yosida theorem are fulfilled.

A.3.1 Cores

We follows here [12]. Let $T(\cdot)$ be a strongly continuous semigroup in X with infinitesimal generator A. A linear subspace Y of $D(A)$ is said to be a *core* for $T(\cdot)$ (or for A) if it is dense in $D(A)$ (endowed with its graph norm). In other words, Y is a core for $T(\cdot)$ (or for A) if and only if $Y \subset D(A)$ and for any $x \in D(A)$ there exists a sequence $(x_n) \subset Y$ such that $x_n \to x$ and $Ax_n \to Ax$ in X.

Proposition A.19 *Let $T(\cdot)$ be a strongly continuous semigroup with infinitesimal generator A, and let $Y \subset D(A)$ be a subspace of X. Assume that*

(i) Y is dense on X.
(ii) $T(t)(Y) \subset Y$ for all $t \geq 0$.

Then Y is a core for T.

Proof. Let $x \in D(A)$. Choose λ larger than the type of $T(\cdot)$. Set $y = \lambda x - Ax$, and denote by (y_k) a sequence in Y such that $y_k \to y$ in X. Set $x_n = R(\lambda, A)y_n$. Then clearly

$$x_n \to x, \quad Ax_n \to Ax \quad \text{in } X.$$

However x_n does not belong necessarily to Y, since Y is invariant for $T(t)$ but not (in general) for $R(\lambda, A)$. Then we consider an approximation $x_{n,\varepsilon}$ of x_n that belongs to Y by setting

$$x_{n,\varepsilon} = \sum_{k=1}^{N_\varepsilon} e^{-\lambda t_\varepsilon} T(t_k^\varepsilon)(t_{k+1}^\varepsilon - t_k^\varepsilon)y_n,$$

where N_ε and $T(t_k^\varepsilon)$, $k = 1, \ldots, N_\varepsilon$, are chosen such that

$$x_{n,\varepsilon} \to x_n, \quad Ax_{n,\varepsilon} \to Ax_n, \text{ as } \varepsilon \to 0.$$

Finally, setting

$$\overline{x_n} = x_{n,\frac{1}{n}},$$

we have $x_{n,\frac{1}{n}} \in Y$ and

$$x_{n,\frac{1}{n}} \to x, \quad Ax_{n,\frac{1}{n}} \to A, \text{ as } n \to \infty.$$

Therefore Y is a core for A. $\square$

A.4 Dissipative operators

Let $A : D(A) \subset X \to X$ be a linear operator. A is said to be *dissipative* if

$$\Re \langle Ax, x \rangle \leq 0 \quad \text{for all } x \in D(A). \tag{A.19}$$

The dissipative operator A is called m-dissipative if $(\lambda - A)(D(A)) = X$, for all λ such that Re $\lambda > 0$. For instance if A is the infinitesimal generator of a contraction semigroup $T(\cdot)$, then it is m-dissipative. We have in fact, for all $x \in D(A)$,

$$\Re \langle Ax, x \rangle = \lim_{h \to 0} \Re (\langle T(h)x - x, x \rangle) \leq 0.$$

A *self-adjoint* operator is an m-dissipative hermitian operator.

Proposition A.20 *Assume that A is m-dissipative. Then A is the infinitesimal generator of a contraction semigroup.*

Proof. Consider the equation

$$\lambda x - Ax = y,$$

where $y \in X$ and $\Re \lambda > 0$. Multiplying both sides of the equation by x and taking the real part we find

$$\Re \lambda |x|^2 + \Re \langle Ax, x \rangle = \Re \langle x, y \rangle.$$

By the dissipativity of A it follows that

$$|x| \leq \frac{1}{\Re \lambda} |y|.$$

Therefore A fulfills the assumptions of the Hille–Yosida theorem and the conclusion follows. $\square$

We prove now an important result due to G. Lumer and R. S. Phillips, see e.g. [23].

Theorem A.21 *Let $A \colon D(A) \subset X \to X$ be a dissipative operator with domain $D(A)$ dense in X. Assume that for any $\lambda > 0$ the range of $\lambda - A$ is dense on X. Then A is m-dissipative.*

Proof. We proceed in two steps.

Step 1. A is closable.

Let $(x_n) \subset D(A)$ be such that

$$x_n \to 0, \quad Ax_n \to y,$$

for some $y \in X$. Since A is dissipative for any $\lambda > 0$ and any $u \in D(A)$ we have

$$|u| \leq \frac{1}{\lambda} |\lambda u - Au|. \tag{A.20}$$

Let now z be any element in $D(A)$. Setting $u = z + \lambda x_n$ in (A.20) yields

$$|z + \lambda x_n| \le \frac{1}{\lambda} |\lambda(z + \lambda x_n) - A(z + \lambda x_n)|,$$

which is equivalent to

$$|z + \lambda x_n| \le |z + \lambda x_n - \frac{1}{\lambda} Az - Ax_n|.$$

Letting n tend to infinity yields

$$|z| \le |z - \frac{1}{\lambda} Az - y|.$$

Now, letting λ tend to ∞, we obtain

$$|z| \le |z - y|,$$

which yields $y = 0$ since one can choose z arbitrarily close to y by the density of $D(A)$.

Let us now denote by $\overline{A}$ the closure of A. It is easy to see that $\overline{A}$ is dissipative.

Step 2. $(\lambda - A)(D(\overline{A}))$ is closed for any $\lambda > 0$.

In fact, let $\lambda > 0$ and let (y_n) be a Cauchy sequence on X. Moreover let $x_n \in D(A)$ be such that

$$\lambda x_n - \overline{A} x_n = y_n, \quad n \in \mathbb{N}.$$

Since $\overline{A}$ is dissipative we have for any $m, n \in \mathbb{N}$

$$|x_n - x_m| \le \frac{1}{\lambda} |y_n - y_m|.$$

Therefore the sequence (x_n) is Cauchy and consequently is convergent to some element $x \in X$. Since $\overline{A}$ is closed $x \in D(\overline{A})$ and the conclusion follows.

Now the theorem is proved since $(\lambda - A)(D(\overline{A}))$ is both closed and dense in X. $\square$

Bibliography

1. P. Billingsley, *Probability and measure*, Third edition. John Wiley & Sons, New York, 1995.
2. L. Breiman, *Probability*, Classics in Applied Mathematics, Siam, vol. 7, 1992.
3. H. Brézis, *Operatéurs maximaux monotones*, North-Holland, 1973.
4. H. Brézis, *Analyse fonctionnelle*, Masson, 1983.
5. A. Chojnowska-Michalik and B. Goldys, *Existence, uniqueness and invariant measures for stochastic semilinear equations*, Probab. Th. Relat. Fields, **102**, 331–356, 1995.
6. G. Da Prato, *An introduction to infinite dimensional analysis*, Appunti, Scuola Normale Superiore, Pisa, 2001.
7. G. Da Prato, *Kolmogorov equations for stochastic PDEs*, Birkhäuser, 2004.
8. G. Da Prato, A. Debussche and B. Goldys, *Invariant measures of non symmetric dissipative stochastic systems*, Probab. Th. Relat. Fields, **123**, 3, 355–380.
9. G. Da Prato and J. Zabczyk, *Stochastic equations in infinite dimensions*, Encyclopedia of Mathematics and its Applications, Cambridge University Press, 1992.
10. G. Da Prato and J. Zabczyk, *Ergodicity for infinite dimensional systems*, London Mathematical Society Lecture Notes, n.229, Cambridge University Press, 1996.
11. G. Da Prato and J. Zabczyk, *Second order partial differential equations in Hilbert spaces*, London Mathematical Society Lecture Notes, n.293, Cambridge University Press, 2002.
12. E. B. Davies, *One parameter semigroups*, Academic Press, 1980.
13. J. D. Deuschel and D. Stroock, *Large deviations*, Academic Press, 1984.
14. N. Dunford and J. T. Schwartz, *Linear operators. Part I*, Interscience, 1964.
15. N. Dunford and J. T. Schwartz, *Linear operators. Part II*, Interscience, 1964.
16. L. Gross, *Logarithmic Sobolev inequalities*, Amer. J. Math., **97**, 1061–1083, 1976.
17. M. Hairer and J. C. Mattingly. *Ergodicity of the 2D Navier–Stokes Equations with Degenerate Stochastic Forcing* to appear on Annals of Mathematics.
18. R. S. Halmos, *Measure theory*, Van Nostrand, 1950.
19. N. Ikeda and S. Watanabe, *Stochastic differential equations and diffusion processes*, North Holland, 1981.
20. R. Z. Khas'minskii, *Stochastic stability of differential Equations*, Sijthoff and Noordhoff, 1980.

21. M. Métivier, *Notions fondamentales de la théorie des probabilitées*, Dunod Université, 1968.
22. K. P. Parthasarathy, *Probability measures in metric spaces*, Academic Press, New York, 1967.
23. A. Pazy, *Semigroups of linear operators and applications to partial differential equations*, Springer-Verlag, 1983.
24. M. Röckner, L^p-*analysis of finite and infinite dimensional diffusions,* Lecture Notes in Mathematics, vol. 1715, Springer, 65-116, 1999.
25. D. Stroock, *Logarithmic Sobolev inequalities for Gibbs states,* Lecture Notes in Mathematics, vol. 1563, Springer, 194–228, 1993.
26. K. Yosida, *Functional analysis,* Springer-Verlag, 1965.
27. J. Zabczyk, *Linear stochastic systems in Hilbert spaces: spectral properties and limit behavior,* Report n. 236, Institute of Mathematics, Polish Academy of Sciences, 1981. Also in Banach Center Publications, **41**, 591–609, 1985.

Index

Universitext

Demazure, M.: Bifurcations and Catastrophes

Devlin, K. J.: Fundamentals of Contemporary Set Theory

DiBenedetto, E.: Degenerate Parabolic Equations

Diener, F.; Diener, M.(Eds.): Nonstandard Analysis in Practice

Dimca, A.: Sheaves in Topology

Dimca, A.: Singularities and Topology of Hypersurfaces

DoCarmo, M. P.: Differential Forms and Applications

Duistermaat, J. J.; Kolk, J. A. C.: Lie Groups

Edwards, R. E.: A Formal Background to Higher Mathematics Ia, and Ib

Edwards, R. E.: A Formal Background to Higher Mathematics IIa, and IIb

Emery, M.: Stochastic Calculus in Manifolds

Emmanouil, I.: Idempotent Matrices over Complex Group Algebras

Endler, O.: Valuation Theory

Erez, B.: Galois Modules in Arithmetic

Everest, G.; Ward, T.: Heights of Polynomials and Entropy in Algebraic Dynamics

Farenick, D. R.: Algebras of Linear Transformations

Foulds, L. R.: Graph Theory Applications

Franke, J.; Härdle, W.; Hafner, C. M.: Statistics of Financial Markets: An Introduction

Frauenthal, J. C.: Mathematical Modeling in Epidemiology

Freitag, E.; Busam, R.: Complex Analysis

Friedman, R.: Algebraic Surfaces and Holomorphic Vector Bundles

Fuks, D. B.; Rokhlin, V. A.: Beginner's Course in Topology

Fuhrmann, P. A.: A Polynomial Approach to Linear Algebra

Gallot, S.; Hulin, D.; Lafontaine, J.: Riemannian Geometry

Gardiner, C. F.: A First Course in Group Theory

Gårding, L.; Tambour, T.: Algebra for Computer Science

Godbillon, C.: Dynamical Systems on Surfaces

Godement, R.: Analysis I, and II

Goldblatt, R.: Orthogonality and Spacetime Geometry

Gouvêa, F. Q.: p-Adic Numbers

Gross, M. et al.: Calabi-Yau Manifolds and Related Geometries

Gustafson, K. E.; Rao, D. K. M.: Numerical Range. The Field of Values of Linear Operators and Matrices

Gustafson, S. J.; Sigal, I. M.: Mathematical Concepts of Quantum Mechanics

Hahn, A. J.: Quadratic Algebras, Clifford Algebras, and Arithmetic Witt Groups

Hájek, P.; Havránek, T.: Mechanizing Hypothesis Formation

Heinonen, J.: Lectures on Analysis on Metric Spaces

Hlawka, E.; Schoißengeier, J.; Taschner, R.: Geometric and Analytic Number Theory

Holmgren, R. A.: A First Course in Discrete Dynamical Systems

Howe, R., Tan, E. Ch.: Non-Abelian Harmonic Analysis

Howes, N. R.: Modern Analysis and Topology

Hsieh, P.-F.; Sibuya, Y. (Eds.): Basic Theory of Ordinary Differential Equations

Humi, M., Miller, W.: Second Course in Ordinary Differential Equations for Scientists and Engineers

Hurwitz, A.; Kritikos, N.: Lectures on Number Theory

Huybrechts, D.: Complex Geometry: An Introduction

Isaev, A.: Introduction to Mathematical Methods in Bioinformatics

Istas, J.: Mathematical Modeling for the Life Sciences

Iversen, B.: Cohomology of Sheaves

Jacod, J.; Protter, P.: Probability Essentials

Jennings, G. A.: Modern Geometry with Applications

Jones, A.; Morris, S. A.; Pearson, K. R.: Abstract Algebra and Famous Inpossibilities